Making of America Project

Easy rules for the measurement of earthworks by means of the

prismoidal formula

Making of America Project

Easy rules for the measurement of earthworks by means of the prismoidal formula

ISBN/EAN: 9783337156633

Printed in Europe, USA, Canada, Australia, Japan

Cover: Foto ©berggeist007 / pixelio.de

More available books at **www.hansebooks.com**

FOR THE

MEASUREMENT OF EARTHWORKS,

BY MEANS OF THE

PRISMOIDAL FORMULA.

ILLUSTRATED WITH NUMEROUS WOODCUTS, PROBLEMS, AND EX-
AMPLES, AND CONCLUDED BY AN EXTENSIVE TABLE
FOR FINDING THE SOLIDITY IN CUBIC
YARDS FROM MEAN AREAS.

THE WHOLE

BEING ADAPTED FOR CONVENIENT USE BY ENGINEERS, SURVEYORS,
CONTRACTORS, AND OTHERS NEEDING CORRECT
MEASUREMENTS OF EARTHWORK.

BY

ELLWOOD MORRIS, CIVIL ENGINEER.

PHILADELPHIA:
T. R. CALLENDER & CO., THIRD AND WALNUT STS.
LONDON: TRÜBNER & CO., 60 PATERNOSTER ROW.
1872.

Dedication.

TABLE OF CONTENTS,

BY CHAPTER, ARTICLE, PAGE, AND REFERENCE TO ILLUSTRATIONS.

———

CHAPTER I.

PRELIMINARY PROBLEMS.

EASY RULES

MEASUREMENT OF EARTHWORKS,

BY MEANS OF THE PRISMOIDAL FORMULA.

CHAPTER I.

PRELIMINARY PROBLEMS.

1. *Of the Prismoid.*—Although this solid probably originated with the ancient geometers—THOMAS SIMPSON (1750), an eminent mathematician of the last century, appears to have been *the first*, in later days, to demonstrate the rule for its solidity,[*] now accepted by modern mensurators; and he was soon followed by Hutton, in his quarto treatise on Mensuration,[†] who by another process again demonstrated the Prismoidal Rule, and at the same time laid the foundations of modern mensuration, in a manner so solid, that it has come down to our time, through various editors and commentators, *substantially* (in many cases literally) *the same* as established by Hutton in his famous work of 1770.

Simpson's rule for the prismoid has been variously transformed, and written, and is now generally known by the name of *the prismoidal formula*, of which we will give hereafter the usual expressions, as well as some useful modifications, the same in substance, but often more convenient for practical purposes.

The solid called a Prismoid (from its general resemblance to a prism, and in like manner named from its base, triangular, rectangular, trapezoidal, etc.) *is a body contained between two parallel planes,*

---

[*] Simpson's Doctrine of Fluxions. (1750), 8vo, London.
[†] Hutton's Mensuration. (1770), 4to, Newcastle upon Tyne.

its hight being their perpendicular distance apart, its ends rectangles, and its faces plane trapezoids;—and this seems to be a sufficient definition. As to such form, *all prismoids* may be reduced or made *equivalent;* but although this simple definition answers our purpose of introducing *the rectangular prismoid,* HUTTON's, *Art. 3, is the authoritative one.*

This solid is usually the frustum of a wedge; but as the proportions of the ends are changed, it may become a frustum of a pyramid, a complete pyramid, a wedge, or a prism; and hence it is indispensably necessary that the rule for its solidity should also hold for *all* these solids, which, in fact, *it does.*

The ends may be, *and often are,* irregular polygons, but they must always coincide with the limiting parallel planes; and though the solid may be quite oblique, its hight must be taken normal to the end planes. The faces are usually straight longitudinally, but this condition is not absolute, since the remarkable formula, deduced from the prismoid for its solidity, applies as well to the volume of many curved solids in an extraordinary manner, of which the limits are not yet known, though more than a century has elapsed since Simpson developed it.

The *mid-section,* included by the usual prismoidal formula, must be in a plane parallel to, and equally distant from, those containing the ends, and is deduced from the arithmetical average of like parts in them. It is entirely hypothetical, or assumed for the purposes of computation, and has no actual existence in the body itself.

The rectangular prismoid (usually regarded as the elementary figure of this solid) is a frustum of the wedge.

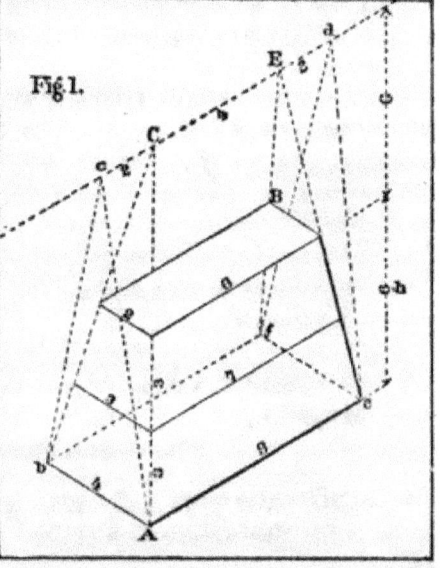

Fig. 1.

(a.) ...... Thus the prismoid AB (*Fig.* 1) is a frustum of the wedge AEC.

The wedge AEC itself being a triangular prism, truncated *twice*, the rectangular prismoid then is a triangular prism, *trebly truncated :* 1st, by two cutting planes, reduced to a wedge; and 2nd, by another plane, to a prismoid (AB), the latter being parallel to the base, and by its section forming the top of the solid at B.

The prismoid, therefore, may be computed as a truncated triangular prism or wedge, and the part cut off deducted, in like manner as the frustum of a pyramid may be calculated as though the pyramid was complete, and then the truncated part computed separately and subtracted, leaving only the solidity of the frustum, subject, like the prismoid, to calculation, by more concise rules, if expedient.

Referring now to *Fig.* 1.

Let A *b c d e f* be the original triangular prism, truncated right and left by planes passing through A *b* and *e f*, reducing it *first* to the wedge AE; and *secondly*, by passing the plane B 2, parallel to the base *c b*, leaving as the residual solid, after three truncations, *the Prismoid* AB.

Then, in the wedge AEC, the right section has a base of 4, a hight of 12, and area of 24, which, multiplied by ⅓ the sum of the lateral edges * (or 6⅔), gives a solidity of 160; while the wedge BCE, *cut off*, has a base of 2, and hight of 6, in its right section, or area of 6, which, multiplied by ⅓ the sum of its lateral edges (or 5⅓), gives a volume of 32.

Now, 160 — 32 = 128, the solidity of the Prismoid AB, as is shown (more concisely) *as follows :*

By Simpson's Rule—

	Hts.	Widths.	
Base, ...	8 × 4	=	32
Top, ...	6 × 2	=	12
Product of sums, equivalent to } 4 times mid. sec., ... }	14 × 6	=	84
			128
Multiplied by ⅙ **h.**		=	1
Solidity, ...		=	128

(The same as above.)

Precisely the same result is also reached by means of the centre of gravity of the right section, flowing with that section along a line

---

* Chauvenet's Geom. (1871), vii. 22. *A wedge*, whether trapezoidal or rectangular, being merely a truncated triangular prism, this rule of Chauvenet's is probably the most concise, and *best for ordinary use.*

curved with an infinite radius, according to Hutton's Problem.* The right section of the prismoid AB (*Fig.* 1) is a plane trapezoid (18 in area), of which (from the dimensions given in the figure) the centre of gravity is found in a perpendicular line, drawn from the middle of A *b*, and at the distance of $2\frac{3}{7}$ feet vertically from it. Now, the length of a straight line, drawn from face to face of the prismoid, parallel to the plane of the base—also to its edges—and at a vertical distance of $2\frac{3}{7}$ feet, will be $7\frac{1}{9}$ feet, by which the right section (18) being multiplied, we have for the *solidity* = 128, as before.

2. THOMAS SIMPSON's *Prismoidal Rule.*—In his work on Fluxions and their Applications (1750), Simpson demonstrates the following rule for the solidity of a prismoid, referring to *Fig.* 2.

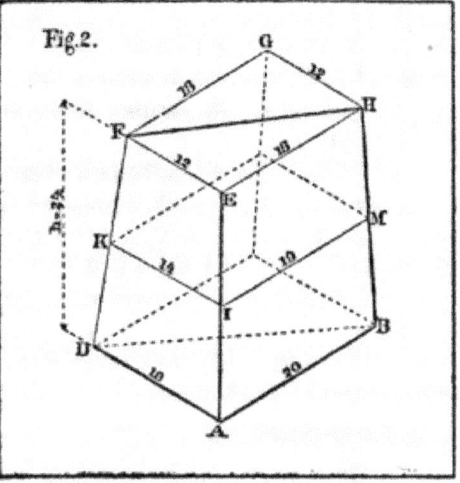

Fig. 2.

This rule for the prismoid, as demonstrated by Simpson, renders the formation of the hypothetical mid-section unnecessary, though containing it, *in effect*, as marked upon the figure, for illustration.

Simpson's Rule is as follows:—Fig. 2.

$$(AB \times AD) + (EH \times EF) + (\overline{AB + EH} \times \overline{AD + EF}) \times$$
$$\tfrac{1}{6}\, \mathbf{h} = Solidity, \quad \ldots \ldots \ldots \ldots \quad \textbf{(I.)}$$

Or,

$$\left(\begin{array}{c}\text{hight} \times \text{width} \\ \text{of one end,}\end{array}\right) + \left(\begin{array}{c}\text{hight} \times \text{width} \\ \text{of other end,}\end{array}\right) +$$
$$\left(\begin{array}{c}\text{sum of hights} \times \text{sum of widths} \\ \text{of both ends,}\end{array}\right) \times \tfrac{1}{6}\, \mathbf{h} = Solidity, \ldots \textbf{(I.)}$$

Here AB × AD = area of base. EH × EF = area of top. While the product of their sums = (AB + EH) × (AD + EF) = four times the area of the mid-section.

EXAMPLE 1.

Let AB and EH be called the *widths*, AD and EF the *hights*, and take the dimensions marked upon *Fig.* 2. Then, by Simpson's rule, we have for the solidity of this *rectangular* prismoid the following:

$$
\begin{aligned}
&\text{Widths.} \quad \text{Hts.} \\
&20 \times 16 = 320 = \text{area of base.} \\
&18 \times 12 = 216 = \quad \text{do. top.}
\end{aligned}
$$

Sums of hts. and widths $= 38 \times 28 = 1064 =$ four times mid-sec.

$$1600 = \text{sum of areas.}$$

Multiplied by $\frac{1}{6} h = \frac{24}{6}$, ... $= \quad 4 = \frac{1}{6} h.$

Solidity, ... $= 6400 =$ volume.

(**a.**) . . . . . The above is a *rectangular prismoid*, or one in which all the parallel sections are rectangles. Now, suppose this prismoid to be cut diagonally by a plane, FHBD, dividing it into two *triangular prismoids*, each equal to the other, and to one-half of the rectangular prismoid.

Then $(AB \times AD) = $ *double the base;* $(EH \times EF) = $ *double the top;* and $(AB + EH) \times (AD + EF) = $ *eight times the mid-section.*

Hence, Simpson's rule, though applicable to any prismoid, by reducing the ends *to equivalent rectangles,* seems especially suitable to triangular prismoids, since the double area of every triangle is equal to the product of its hight and width, taken rectangularly; while the product of the sums of those hights and widths, multiplied together, gives eight times the area of the mid-section, without the necessity of forming it by arithmetical averages.

Accordingly, with triangular sections, a slight transformation of this rule will often be more convenient for use *with given areas.*

Thus,

Let *double* the area of the base . . . . $= 2 \mathbf{b.}$
"   "   "   top . . . . $= 2 \mathbf{t.}$
Eight times the area of the mid-sec. . . . . . $= 8 \mathbf{m.}$
And the final divisor (12), or if used as above, . $= \frac{1}{12} \mathbf{h.}$

Then, to find, in the first instance, *the mean area* of the prismoid.

We have the formula, $\dfrac{2 \mathbf{b} + 2 \mathbf{t} + 8 \mathbf{m}}{12} = $ *mean area* . . (**II.**)

And this mean area, being multiplied by the hight or length ($\mathbf{h}$), of the whole prismoid between the end planes, gives *the solidity.*

Thus, in the case of *the two triangular* prismoids, into which the diagonal plane FB (*Fig.* 2) divides Simpson's *rectangular* prismoid, we have, by taking the dimensions marked upon the figure,—*the following:*

EXAMPLE 2.

Calculation of the triangular prismoid ABDFHE, or of its equal GD = 3200, *Solidity.*

$$
\begin{array}{llll}
\text{Hts.} & \text{Widths.} & & \\
16 \times 20 & = & 320 & = 2\ \mathbf{b.} \\
12 \times 18 & = & 216 & = 2\ \mathbf{t.} \\
\hline
\text{Sums, . . } 28 \times 38 & = & 1064 & = 8\ \mathbf{m.} \\
\hline
& 12)\overline{1600} & & \\
\text{Mean area, . . } & = & 133\tfrac{1}{3} \times \mathbf{h} & = 24 = 3200,\ Solidity.
\end{array}
$$

And $3200 \times 2 = 6400 =$ the solidity of the whole rectangular prismoid, as above.

3. CHARLES HUTTON's *Prismoidal Rules.*—In his famous quarto Mensuration (Newcastle-upon-Tyne, 1770), Hutton gives the following definition:

"A prismoid is a solid having for its two ends any dissimilar parallel plane figures of the same number of sides, and all the sides of the solid, plane figures also."

He adds: "It is evident that the sides of this solid are all trapezoids;" and: "If the ends of the prismoid be bounded by curves, as ellipses, etc., the number of its sides, or trapezoids, will be infinite, and it is then called, sometimes, a cylindroid."

Hutton gives two rules for the solidity of the body (so defined), one *general*, and the other he calls the *particular* rule—he also indicates a third, by means of initial prismoids, which, by a little development, *can be made quite useful.*

Hutton's General Rule.

"To the sum of the areas of the two ends add four times the area of a section parallel to, and equally distant from, both ends, multiply the last sum by the hight, and ⅙ of the product will be *the solidity,* . . . . **(III.)**

In this shape, and nearly in the same words, through Bonnycastle, and other writers on Mensuration, *the Prismoidal Formula* has come down to our time.

In the work above cited, Hutton also (part iv. prop. 3) shows that

¼ of the sum of the end areas, and four times the mid-section, gives *the mean area* of any prismoidal solid, which, multiplied by its length, will equal *the solidity.*

The *particular rule,* referred to above, is directly deduced from that given by him for the solidity of a wedge.

Thus, referring to *Fig.* 3 (copied by us from the original work of 1770).

Hutton says, where L and *l* represent two corresponding dimensions of the end rectangles, B and *b* the others, and **h** the hight or length of the prismoid,

Then,

$$(\overline{2L + l} \times B + \overline{2l + L} \times b) \times \tfrac{1}{6}\,h = \textit{Solidity,}$$

—*which is the particular rule,* . . . . (**IV.**)

A note, on page 163, referring to this, says:

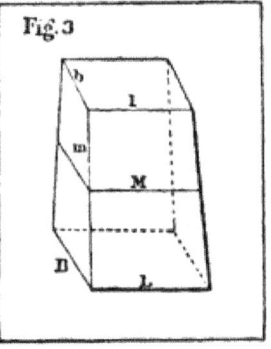

Fig. 3

"It is evident that the rectangular prismoid is composed of two wedges, whose bases are the two ends of the prismoid, and whose hights are each equal to that of the prismoid."

It might be added, that the edges of these two wedges are formed by two diagonally opposite sides of the rectangular ends.

Hutton notes also,

That $\dfrac{L + l}{2} = M$, and $\dfrac{B + b}{2} = m$, the sides of the mid-section, so that the correspondence of the General and Particular Rules becomes evident.

(**a.**) . . . . . . At page 164 of the quarto Mensuration, cited above, reference is made to the General Rule as follows:

"This rule will serve for any prismoid or cylindroid, of whatever figure the ends may be, inasmuch as they may be conceived to be composed of an infinite number of rectangular prismoids. Which is the General Rule."

This method of considering any prismoid to be composed of a great number of rectangular prismoids, of the same common length, has prevailed from Hutton's time down to the present day.

Thus, we find in Davies Legendre,* chapter on the Mensuration

---

* Davies Legendre. (1853), 8vo : New York.

of Solids, in treating of prismoids, where he copies Hutton's figure, and both Particular and General Rules,—*the following :*

"This rule (*the general one*) may be applied to any prismoid whatever. For whatever the form of the bases, there may be inscribed in each the same number of rectangles, and the number of these rectangles may be made so great that their sum in each base will differ from that base by less than any assignable quantity. Now, if on these rectangles rectangular prismoids be constructed, their sum will differ from the given prismoid by less than any assignable quantity. Hence, the rule is general."

In his remarkable chapter on the cubature of curves (Mens., part iv. page 457), Hutton shows that the prismoidal formula is applicable to the frusta of all solids generated by the revolution of a conic section (as well as to the complete solids); also, to all pyramids and cones, and in short to all solids (right or oblique), *of which the parallel sections are similar figures.*

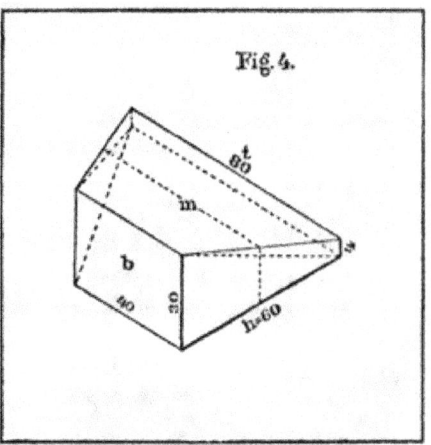

Fig. 4.

We will now illustrate Hutton's Rules, by means of a figure and examples, to find the solidity of a prismoid, *with very dissimilar ends.* (See *Fig. 4.*)

1. *By General Rule.*

$$40 \times 30 = 1200 = \mathbf{b}.$$
$$80 \times 4 = 320 = \mathbf{t}.$$
$$60 \times 17 \times 4 = 4080 = 4\,\mathbf{m}.$$
$$6)\overline{5600}$$
$$933\tfrac{1}{3}$$

Multiplied by $\mathbf{h} =$ 60

Solidity = 56000

2. *By Particular Rule.*

As two Wedges.

40	80
2	2
80	160
80	40
160	200
30	4
4800	800

$\tfrac{1}{6}\mathbf{h}$ . .   10   10

| 48000 | 8000 |
| 8000 | |

Solidity = 56000 of whole prismoid.

3. *By means of Initial Prismoids.* . . . . . . (**V.**) (To be further explained.)

(1) Areas of ends, **b** = 1200, and **t** = 320.

(2) $\left\{\begin{array}{l}\text{Hights} = 30\\\text{Widths} = 40\end{array}\right\}$ **b** $= \left.\begin{array}{l}4\\80\end{array}\right\}$ **t.**

(3) Assumed squares in larger end, 1200 of 1 × 1.

(4) Ratio of ends, $\dfrac{t}{b} = \dfrac{320}{1200} = \cdot 2667.$

(5) Proportional rectangles in small end (1200 in number), $\dfrac{80}{40} = 2,$

$\dfrac{4}{30} = \cdot 13333,\ 2 \times \cdot 13333 = \cdot 26667 =$ area of these, being equivalent to the ratio of the ends 1 to $\cdot 2667.$ [See (4).]

(6) *Mid-section,* dimensions of proportional rectangle, $\dfrac{1+2}{2} = 1\cdot 5,$

$\dfrac{1 + \cdot 13333}{2} = \cdot 5667,$ and $1\cdot 5 \times \cdot 5667 = \cdot 85 =$ rectangular area of mid-section of initial prismoid.

(7) $\left\{\begin{array}{l}\text{\quad Then for the solidity of the initial prismoid, by General}\\\text{Rule.}\\\text{\quad Call these areas}\\\text{\textbf{b}}', \text{\textbf{m}}', \text{and \textbf{t}}', \text{to}\\\text{distinguish them}\\\text{from those of the}\\\text{main solid.}\end{array}\right.$

$\left.\begin{array}{l}\textbf{b}' = 1 \times 1 \ . \ . = 1\\4\,\textbf{m}' = \cdot 85 \times 4 \ . = 3\cdot 4\\\textbf{t}' = \cdot 13333 \times 2 = \quad \cdot 26667\\\hline\qquad\qquad\qquad 6)\ 4\cdot 66667\end{array}\right.$

Mean area, ... = $\cdot 77778$

Multiplied by **h** . . = 60

Volume of one, ... = 46·66680

Mult. by No. initial prismoids, assumed = 1200

Solidity of the whole prismoid, as above = 56000·16000

In computing initial prismoids it is necessary to employ sufficient decimals, but 4 or 5 places *are usually enough.*

No. of initial prismoids assumed = 1200.

(**b.**). . . . . These initial prismoids are supposed to be constructed upon small rectangles in the two ends, *equal in number in each, and of proportional areas.*

In the base, or larger end (though either end may be used), it will be most convenient to assume these to be *squares* formed upon the unit of measure, while at the top *they must be rectangles proportional both in dimensions and area,* by the view we have herein taken (as indicated at (5) above).

The end areas of the main prismoid being always given, or computable, they must be proximately reduced to rectangles before we can properly apply the principle of initial prismoids to calculate, or verify, their solidity;—and the solid will then become, in effect, a rectangular prismoid like those of Simpson and Hutton.

In doing this, it will be sufficient to dermine a width and hight, apparently proportional to the shape of the cross section (which in some species of earthwork is extremely irregular),—but this hight and width must be such that, used as factors, they reproduce the given area, even though of themselves they may not be *exactly* geometrical equivalents, for the dimensions of the section.

Having thus (as it were) rectified the solid proximately, we may proceed with it as a rectangular prismoid, *by the method of initial prismoids*, briefly as follows:—*Determine the rectangular hights and widths, such as will proximate the figure, and by multiplication reproduce the areas. Assume one end as base, to be divided into squares of superficial units, and the others into proportional rectangles; upon these construct (or imagine) initial prismoids, and having ascertained the volume of one, multiply by number, for solidity of main prismoid, as shown in detail above. . . .* (V.)

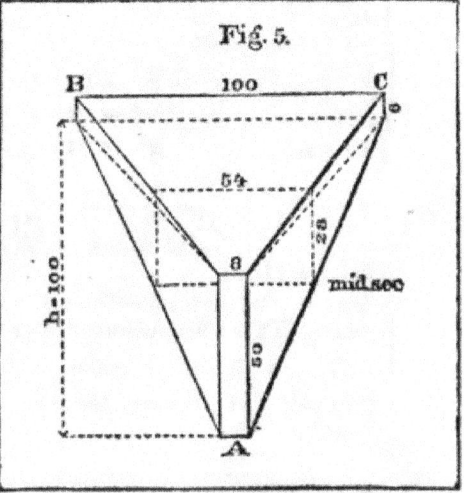

Fig. 5.

(c.) . . . . . . We will further illustrate this subject by presenting an outline of a T-shaped prismoid; a solid (*Fig. 5*), with a figure so peculiar that none of the usual methods of averaging could even proximate its solidity, ... which can only be dealt with by *the Prismoidal Formula*, or some cognate rules.

This we will calculate as a prismoid by Simpson's General Rule, by Hutton's Particular Rule, and *by the Method of Initial Prismoids*.

By Hutton's Particular Rule.		By Simpson's General Rule.	
As two Wedges.		*As a Rectangular Prismoid.*	

<table>
<tr><td colspan="2" align="center">By Hutton's Particular Rule.</td><td colspan="2" align="center">By Simpson's General Rule.</td></tr>
<tr><td colspan="2" align="center">As two Wedges.</td><td colspan="2" align="center">As a Rectangular Prismoid.</td></tr>
<tr><td align="center">100</td><td align="center">8</td><td colspan="2" align="center">Hts. Wds.</td></tr>
<tr><td align="center">2</td><td align="center">2</td><td colspan="2" align="center">6×100 . . $=$ 600</td></tr>
<tr><td align="center">——</td><td align="center">——</td><td colspan="2" align="center">50×8 . . $=$ 400</td></tr>
<tr><td align="center">200</td><td align="center">16</td><td colspan="2" align="center">Sums, $56 \times 108 =$</td></tr>
<tr><td align="center">8</td><td align="center">100</td><td colspan="2" align="center"></td></tr>
<tr><td align="center">——</td><td align="center">——</td><td colspan="2" align="center">4 times mid-sec. $=$ 6048</td></tr>
<tr><td align="center">208</td><td align="center">116</td><td colspan="2" align="center">————</td></tr>
<tr><td align="center">6</td><td align="center">50</td><td colspan="2" align="center">7048</td></tr>
<tr><td align="center">——</td><td align="center">——</td><td colspan="2" align="center"></td></tr>
<tr><td align="center">1248</td><td align="center">5800</td><td colspan="2" align="center">$\frac{1}{2}$ h . . . . . $=$ 16$\frac{2}{3}$</td></tr>
<tr><td align="center">100</td><td align="center">100</td><td colspan="2" align="center">Solidity,. . . $= 117466\frac{2}{3}$</td></tr>
<tr><td align="center">6) 124800</td><td align="center">6) 580000</td><td colspan="2" align="center">——————</td></tr>
<tr><td align="center">————</td><td align="center">————</td><td colspan="2"></td></tr>
<tr><td align="center">20800</td><td align="center">96666$\frac{2}{3}$</td><td colspan="2"></td></tr>
<tr><td align="center"></td><td align="center">20800</td><td colspan="2"></td></tr>
<tr><td align="center"></td><td align="center">————</td><td colspan="2"></td></tr>
<tr><td colspan="2" align="center">Solidity $= 117466\frac{2}{3}$</td><td colspan="2"></td></tr>
</table>

By the Method of Initial Prismoids.—Let their number be 400, the same as the superficies of A. Suppose them constructed upon squares at A. (on a side equal to the unit of measure), and upon proportional rectangles at BC.

Then, $600 \div 400 = 1\cdot5$, the ratio of A. to BC. and of initial squares at one end to rectangles at the other.

And in the 3 main sections of the prismoidal solid, *Fig.* 5, We have for similar sections of the initial prismoids $=$

Representative.	Dimensions of initial sections.	Initial areas.	No.	Main areas.
End A . . . $=$ squares of 1×1 . . . . $= 1\cdot$			$\times 400 =$	400.
" BC . . $=$ propor. rectans. $12\cdot5 \times \cdot12 = 1\cdot5$			$\times 400 =$	600.
Mid-section . $=$ " " $6\cdot75 \times \cdot56 = 3\cdot78$			$\times 400 =$	1512.

It will be seen that the main areas result as above calculated ;—*and having these and the common length* **h**, it is easy to compute the prismoid by Simpson's General Rule, as shown before.

We may add here, as being indicative of the difficulty of computing such a solid, by ordinary average rules (which answer tolerably well), in common cases.

That the Arithmetical Mean of the end areas $=$ 500, the Geometrical Mean $=$ 490 ; while the Prismoidal Mid-section $=$ 1512, and the Prismoidal Mean Area $= 1174\frac{2}{3}$; which, multiplied by the length, or hight, **h** $=$ 100 : makes the *solidity*, above $= 117466\frac{2}{3}$, or more than *twice as much* as would result from multiplying the arithmetical mean by the length.

2

4. *The Prismoid adapted to Earthwork.*—Sir John Macneill, a distinguished English engineer, as early as 1833, soon after the introduction of railroads, when the necessity became apparent of having ready and correct methods at hand for computing the volume of the vast quantities of earth, removed or supplied, in grading them, prepared and published three series of Tables (in 8vo), computed by means of the *Prismoidal Formula.* These Tables were systematically arranged, and have been extensively used abroad.

He considered the Earthwork Prismoid *as being composed of a Prism, with a wedge superposed :* since the lower portion of the cross section of a railroad, canal, or road is generally symmetrical and regular, the ground surface alone being relatively variable.

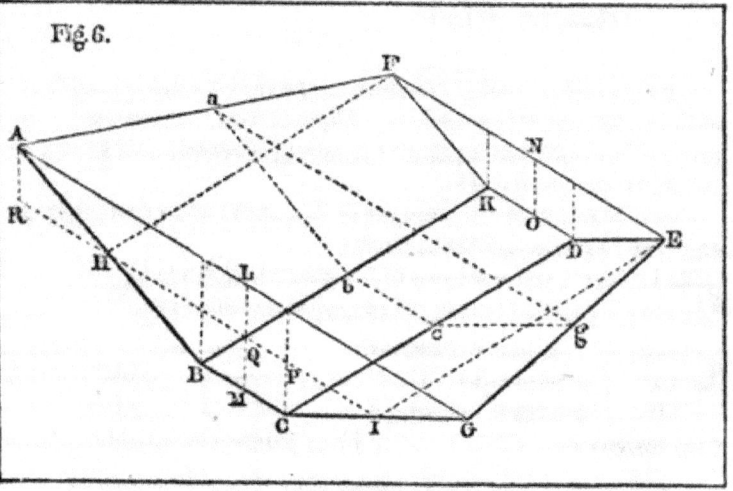

Fig. 6.

In this diagram (*Fig.* 6) the reduced surface of the ground (taken as level, crosswise, or made so) is shown by the plane AFGE, and the cross section of the road by ABCG, these are supposed to be *transparent,* in order to show the road-bed and mid-section, as well as the far end of the trapezoidal prismoid.

Sir John Macneill commences his work, by referring to a representation of the Earthwork Prismoid (*copied above*), as follows :

" Let ABCGFKDE represent a prismoid or solid figure, similar to that which is formed in excavations or embankments, in which BCDK represents the roadway, and ABCG, FKDE, parallel cross sections at each end. The cubic content of this solid is equal to

The area ABCG + area FKDE + 4 times area $abcg$,

$$\text{Mutiplied by } \frac{CD}{6}:$$

"If, then, we suppose a plane, IIIEF, to be drawn through the lines HI, and EF, it will be parallel to the base BCKD, and will divide the solid, ABCGFKDE, into two others, one of which will be the regular prism, HBCIFKDE, and the other will be a wedge, the base of which will be the trapezium, AHIG, the length IE or CD, the length of the prismoid, and the edge FE, the breadth of the cutting at the lower end of the section."

The prismoid, then, being assumed as composed of a regular prism, with a wedge superposed, he demonstrates in the usual manner the formula for the volume of these two solids, and shows that by addition they result in *the Prismoidal Formula,* which he uses in the computation of the three series of Tables which form the bulk of his neat octavo volume (London, 1833).

It will be observed that all Macneill's prismoids refer to ground sloping longitudinally, but *level transversely:*—to apply them, therefore, to an irregular surface, it must be first reduced to a level crosswise, or assumed to be so, *practically.*

The above extract from Sir John Macneill's work of 1833 is made, not only for its intrinsic value, but on account of its being the first regular and successful attempt to adapt *the Prismoidal Formula* to the computation of modern earthworks: which is followed out through a series of practical Tables, comprising 239 pages, and extending to 50 feet of hight or depth:—an embankment being considered as an excavation inverted.

This meritorious work of Sir John Macneill was speedily followed by other writers in England, and later by several in this country.* All, or most of these productions being based upon *the Prismoidal Formula* (or some modification of it), which is now universally acknowledged to be the only consistent and exact method for computing the volume of solids employed in modern earthworks, and even those authors who employ *pyramidal rules* are but using a particular case of the former.

---

* Bidder, Baker, Bashforth, Henderson, Sibley, Rutherford, Hughes, Huntington, Law, Dempsey, Haskoll, Morrison, Rankine, Graham, Macgregor, and others, in England. While in this country, Long, Johnson, Borden, Trautwine, Gillespie, Henck, Davies, P. Lyon, Cross, M. E. Lyons, Byrne, Warner, Rice, and others (besides the present writer), have dealt with this subject. Amongst these, however, the most comprehensive, and the best in many particulars, is the work of John Warner, A. M., a well printed and handsomely illustrated 8vo, Philadelphia, 1861, containing 28 valuable and useful Tables, and 14 plates of great importance to every student of engineering.

5. *The Prismoid in its Simplest Form.*—The unexpected manner in which the Prismoidal Formula applies to the cubature of other solids, totally·dissimilar in form and appearance (as to *the sphere*, taking the poles as end sections at zero, and the mid-section as a great circle), justifies its consideration under various aspects, which would be superfluous in any other body, and hence we give below a figure illustrating the Prismoid, in what may be deemed *its simplest form* (when not contained within a diedral angle). See *Fig.* 7, where the solid is level transversely, but sloping longitudinally, and may be supposed to represent (*proximately*) one of Hutton's *Initial Prismoids*, square at one end, and with a proportional rectangle at the other.

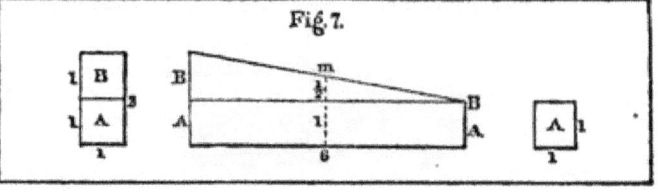

Fig. 7.

Here the prismoid is composed of *a prism* on a square base, with a side of 1, and length of 6,—and of *a wedge*, superposed, with a square back, on a side of 1, its edge also 1, and hight 6,—the common length of the two combined as a prismoid.

$$\text{Let} \begin{cases} \text{A A Represent the prism.} \\ \text{BB The wedge.} \\ m \quad \text{The mid-section of the prismoid.} \end{cases}$$

Then we have for the volume of this solid, by several of the rules already given.

Formulas.   Cubic ft.

$$
\begin{aligned}
&\textbf{(I.)} \quad (1 \times 2) + (1 \times 1) + \left[(1 + 1) \times (2 + 1) \right] \times \frac{6}{6} = 9 \\
&\textbf{(II.)} \quad (2 \times 2) + (2 \times 1) + (1\cdot5 \times 1 \times 8) \div 12 \times 6 \;.\;\; . = 9 \\
&\textbf{(III.)} \quad 2 + 1 + (1\cdot5 \times 1 \times 4) \times \frac{6}{6} \;.\;\; .\;\; .\;\; .\;\; .\;\; .\;\; . = 9 \\
&\textbf{(IV.)} \quad (\overline{2 \times 1 + 1} \times 2) + (\overline{2 \times 1 + 1} \times 1) \times \frac{6}{6} \;.\;\; . = 9 \\
&\quad \begin{array}{l} \text{Divided} \\ \text{Prismoid} \\ \text{of } \textit{Fig. 7.} \end{array} \begin{cases} \text{Prism} \;=\; 1 \times 1 \times 6 \;.\;\; .\;\; .\;\; .\;\; .\;\; . = 6 \\ \text{Wedge} \;=\; \left(\frac{1}{2} \times 6 \times \dfrac{1 + 1 + 1}{3} \right) .\;\; . = 3 \end{cases} = 9
\end{aligned}
$$

All, of course, resulting in the same *solidity* for this simple pris-moid = 9 cubic feet.

6. *Further Illustration of Macneill's Prismoid.*—In computing the quantities of earthwork for railroads, etc., it is often useful (and generally desirable) to consider the side slopes, continued to their intersection, above or below the road-bed (as has been done by T. Baker, C. E.,[*] and other writers), thus forming a constant triangle at the intersection, which is deductive from the general triangular figure formed by the slopes, and ground, in order to obtain the regular cross section of excavation or embankment, from ground to grade; and this triangle also forms the right section of *the grade prism*, terminating the earthwork solid at edge of diedral angle, formed by the side slope planes containing it.

To explain this more clearly, we give a figure in which both end areas are drawn upon the same plane (*Fig.* 8).

Double cross section of a railroad cut—(in fact, Macneill's prismoid on level ground)—with road-bed of 20, and slopes of 1 to 1.

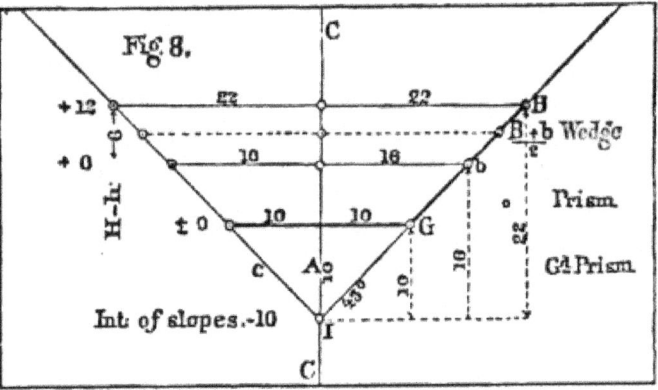

References.

A = Altitude of grade triangle.
B = Level top, sloping forward in 100 feet to *b*.
b = Level top of forward cross section.
G = Grade, or road-bed, 20 feet wide.
c = Grade triangle, or constant end, of grade prism.
H — *h* = Breadth of back of trapezoidal wedge.
r = Slope ratio, or in this case 1.

---

[*] Railway Engineering and Earthwork, by T. Baker, C. E. London, 1840. Wherein he develops a very compendious and excellent system of computing the earthwork of railways, which has been extensively copied.

CC = Centre line of road.

I = Intersection of side slopes, or edge of diedral angle formed by them.

> To find the *equivalent level hight*—no matter how irregular the ground may be.
> Let
>
> a = Whole area, to the intersection of slopes.
> r = Slope ratio.
> h = Equivalent level hight.
>
> Then, $\sqrt{\dfrac{a}{r}} = h$.

Let B and b represent the level tops of *two* cross sections of a railroad cut, 100 feet apart sections, and lying within the same diedral angle of 90°, formed by side slopes of 1 to 1, continued to their intersection, or edge at I.

Now, supposing B and b, to have been originally a very irregular surface, reduced, by any *exact method*, to the level tops represented.

Then, *below b* we have a regular prism, on a triangular base, extending down to I ; and *above b*, a regular wedge (*back and edge parallel*), upon a trapezoidal back, of which the base b is equal to the edge b, representing the top of the forward cross section, 100 feet distant.

Then, in the wedge *above b*, by the properties of that solid, *considered as* * *a truncated triangular prism, and applicable either to rectangular or trapezoidal wedges*,

  We have,

$$\frac{(B + b + b) \times (H - h)}{6} = \frac{(44 + 32 + 32) \times (22 - 16)}{6} = 108. \quad \text{\tiny Mean Area.}$$

And in the prism *below b*, down to I (including the grade triangle)—

  We have,

$$
\left.
\begin{array}{l}
(h^2 r) \quad \cdots \cdots \cdots \cdots \cdots = 256. \\
\text{Deduct the grade triangle} \quad \cdots \cdots = 100.
\end{array}
\right\} \cdot \cdot = 156.
$$

Leaves area of prism (*above grade*) from G to $b = 156.$

Finally, then, we have *the mean area* of the trapezoidal earthwork solid, *above* grade, or road-bed . . . . = 264.

                  Cubic Ft.

Then, $264 \times 100 = 26400.$ *The solidity* of this Prismoid.

---

* Chauvenet's Geom., vii. 22 (1871), easily reducible to the text.

If more convenient, we might exclude entirely the grade triangle, and stop the calculation at G (the road-bed), but as a system of computation, and in view of the simplicity of the geometrical relations of triangles, it will usually be found best to include the grade triangle as above, and ultimately to deduct it, in some form.

The employment of the method of this article enables us to find a mean area to the prismoid—without using a mid-section—and this mean area, when multiplied by the length, gives the volume of the whole solid.

Thus we may assume any level trapezoidal prismoid of unequal parallel ends (as Macneill does), to be composed of two solids—*a prism, with a wedge superposed.*

1. *A Triangular Prism*, with a cross section, equivalent to the lesser end, supposing the slopes to intersect, and embracing the grade triangle.

2. *A Trapezoidal Wedge*, superposed upon the prism, having an area of back equivalent to the difference of the ends, its edge being the level top of the smaller, and equal to the base of the back.

The length being common to both partial solids, and to the whole prismoid.

Then, for the mean area of the wedge, we have,

$$\frac{(\mathrm{B} + b + b) \times (\mathrm{H} - h)^{*}}{6},$$

and for that of the prism to intersection of slopes $= (h^{2} r -$ grade triangle), and by addition,[†]

$$\frac{(\mathrm{B} + b + b) \times (\mathrm{H} - h)}{6} + (h^{2} r - \text{grade triangle}) \times$$

the common length $= The\ Solidity\ of\ the\ Prismoid$ . . . . (**VI.**)

Or, in words,—*The sum of the mean areas of the prism, and superposed wedge, multiplied by the common length, equals the solidity of this prismoid.*

---

* Chauvenet's Geom., vii. 22 (1871).

† B and b are always the widths between top slopes at the ends.

And H — h (however irregular the ground line of the ends may be) is obtained by dividing the difference of end areas by half the sum of their top widths, or $\left(\dfrac{\mathrm{B} + b}{2}\right)$. See note at foot of this **Article 6.**

Note.—When the ground surface, or upper side of the superposed wedge, *is very irregular* (as in *Figs.* 43 and 44)— ascertain the horizontal widths of each end at top slope. Then the difference between the areas of the two ends is the surface of the back of the superposed wedge, and this, divided by the average of the two horizontal widths above, gives the vertical hight of the back, or altitude of the triangular section, of which the length of the prismoid is the base, giving at once the means of computing its area, and this, multiplied by one-third of the sum of the lateral edges, *gives the solidity of the superposed wedge.* (*Chauvenet*, Geom., vii. 22.)

7. *Trapezoidal Prismoid of Earthwork, considered as two Wedges.*— On ground, either level crosswise, or reduced to an equivalent level by any correct process, an Earthwork Prismoid, within the limits of its slopes, road-bed, and ground surface, may readily be computed as *two wedges* (Hutton's Particular Rule), without an assumed mid-section, or even the end areas.

And in this there is some advantage, as the width of road-bed at the end sections may be *unequal* to any extent, provided the widening is gradual.

Thus, let *Fig.* 9 represent a regular station of a railroad cut, 100 feet in length, with slopes of 1 to 1, and in the near end section a depth of 40 feet, and road-bed of 20, while in the far one it has a depth of 30, and road-bed of 40 feet wide.

Hutton's Particular Rule, *modified* for application to earthwork, may be expressed in words at length as follows:

Rule.

In 1st cross section $\left\{\begin{array}{l}\text{Add road-bed + top width + road-}\\\text{bed of 2d section; multiply the sum}\\\text{of these three by level hight of sec-}\\\text{tion, and reserve the product.}\end{array}\right.$

In 2d cross section $\left\{\begin{array}{l}\text{Add road-bed + top width + top}\\\text{width of 1st section; multiply the sum}\\\text{of these three by level hight of sec-}\\\text{tion, and reserve the product.}\end{array}\right.$

Finally, add the two products reserved, and ¼ of their sum is the mean area of the Prismoid, which, multiplied by length = *Solidity.* . . . . **(VII.)**

Referring to *Fig.* 9, the line CC is the centre line traced upon the ground, and below it the road-bed gradually widened from 20 to 40 feet, in the length of 100; the figures marked show the dimensions assumed for illustration, and the dotted lines the edges of a plane supposed to be passed, so as to convert this solid into *two wedges*.

The *nearest* having a trapezoidal *back*, standing on a road-bed of 20, with a hight of 40, and its *edge* being the road-bed of 40 feet wide, belonging to the far cross section.

The *farthest* wedge, above the dotted lines, having for its *back* the

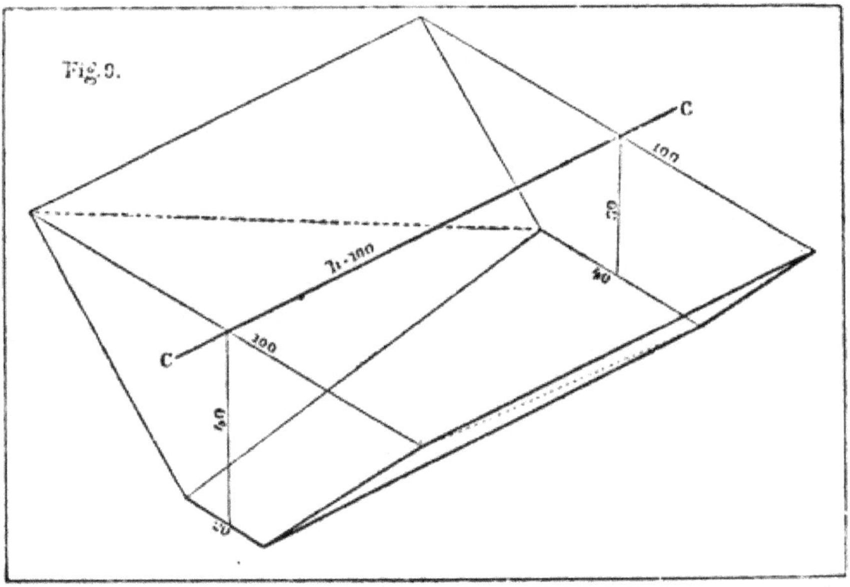

Fig. 9.

far section, standing on a road-bed of 40, with hight of 30, and its *edge* being the top-width of the near cross section, 100 feet wide, *at ground line.*

[In Chapter 5 we shall consider further, and more in detail, the subject of *Wedges;* and their application to the computation of earthwork solids, and illustrate it by several examples. Comparing also the results obtained with those derived from the use of HUTTON'S *General Rule:*—which is the accepted standard for accuracy in such work.]

EXAMPLE.

By Our Modification of Hutton's Rule . . . . . . (**VII.**)	*By Hutton's Particular Rule.* (**IV.**) Reducing Trapezoids to Rectangles.

<table>
<tr><td rowspan="6">In 1st cross section</td><td>20</td><td colspan="2">Mean breadths = 60 = 70</td></tr>
<tr><td>100</td><td>2</td><td>2</td></tr>
<tr><td>40</td><td>120</td><td>140</td></tr>
<tr><td>160</td><td>40</td><td>100</td></tr>
<tr><td>40</td><td>160</td><td>240</td></tr>
<tr><td>6400</td><td>40</td><td>30</td></tr>
<tr><td rowspan="5">In 2d cross section</td><td>40</td><td>6400</td><td>7200</td></tr>
<tr><td>100</td><td colspan="2"></td></tr>
<tr><td>100</td><td colspan="2">6400</td></tr>
<tr><td>240</td><td colspan="2">7200</td></tr>
<tr><td>30</td><td colspan="2">13600</td></tr>
<tr><td></td><td>7200</td><td colspan="2">100</td></tr>
</table>

Finally

$$\begin{array}{r} 6400 \\ 7200 \\ \hline 6)13600 \\ \text{Mean Area} = 2266\cdot67 \\ 100 \\ \hline \text{Solidity} . . = 226667\cdot00 \end{array}$$

$$\begin{array}{r} 6)1360000 \\ \hline \text{Solidity} . . = 226667 \end{array}$$

8. *Areas of Railroad Cross-sections (within Diedral Angles)— whether Triangular, Quadrangular, or Irregular.*

All railroad sections *are contained within diedral angles,* formed by side slope planes, of a given divergency—determined by the slope ratio (r).—The edge of this diedral angle is a right line, parallel to the grade, and prolonged forward indefinitely from I, the intersection of the side slopes (in a right section), until the end of the cut or fill is attained. Here, at the grade point, it changes its position to a corresponding parallel above, or below, as the case may be. Considering, with Sir John Macneill, an embankment to be, in effect, an excavation inverted, the situation of the edge of the diedral angle, or intersection of the slopes, will generally (in our examples) be found *below* the road-bed, but always parallel to the grade line, and at the same distance from it, as long as the side slopes continue uniform.

(**a.**) . . . . . . From the geometrical relations of triangles and rectangles, it is obvious that in a triangle situated as in *Fig.* 10—contained within rectangular axes and their parallels, and divided into two by the central axis h, the area of the whole is equivalent to $\frac{h\,w}{2}$.

— the parallels a and b, to the centre line h, limiting the triangle *laterally*.

The same rule, precisely, applies to quadrangles, which may always be cut by a diagonal into two triangles.

This rule (*in fact*), equally applicable both to triangles and trapeziums, is that laid down by Hutton (1770) for *trapeziums*.

In *Fig*. 10,—$h \times w = double\ area$ of the whole triangle, whose vertex is at I, the intersection of the slopes, and its sides, the side-slopes, and the ground line. Thus, let $h = 20$, $w = 45$, then $20 \times 45 = 900 \div 2 = 450$, area of whole triangle; but it is often more conve-

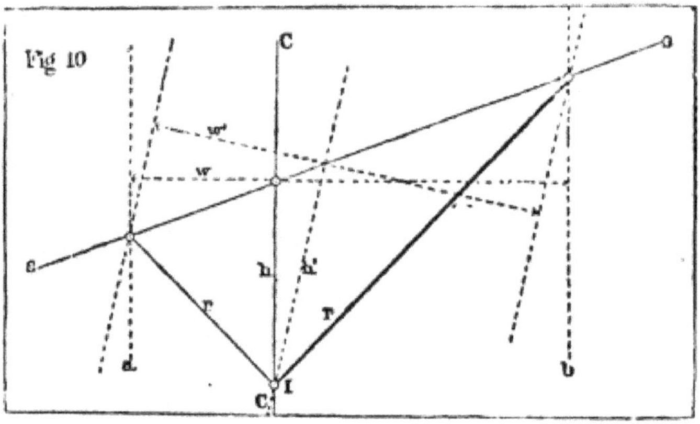

Fig 10

nient, in calculations, to use *double areas alone*, until the close of the operation, as in many problems of land surveying.

In *a triangle*, the direct axes h or h' may take any position, provided the parallels through the lateral vertices are made to follow, and the tranverse axes, w and w', remain rectangular.

But in *a quadrangle*, the position of the direct axis is fixed by that of the opposite vertices, through which it passes, and with it the axis of width, and its limiting parallels, *are also fixed*.

In *Fig*. 10, suppose the direct axis and its parallels to revolve upon I, into the position h', and that h' becomes $22{\cdot}1$—then it will be found that w' has become $40{\cdot}73$, and then, $\dfrac{h' \times w'}{2}$ will be $\dfrac{22{\cdot}1 \times 40{\cdot}73}{2} = 450$, area of whole triangle, *as before*.

In both these cases, *Figs.* 10 and 11, each figure is divided by the centre line, or direct axis, into two triangles, having a common base, and contained between parallels to it, drawn through the opposite vertices.

In both *Figs.* 10 and 11, $h \times w =$ double area of the figure to which they relate,—as these are rectangular factors, for determining the content of the wholly or partially circumscribing rectangles (between the same parallels), of which the triangle or trapezium represented, is each equivalent to *one-half.*

This rule is, in fact, the simplest possible, being, substantially, the definition of a plane surface, length $\times$ breadth (which indicates superficial extension), and from its extreme simplicity, there seems to

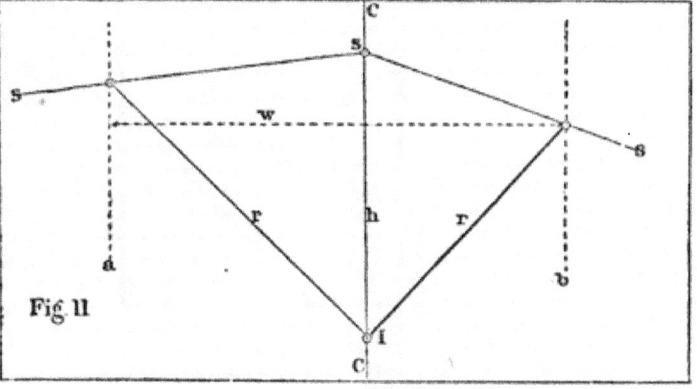

Fig 11

be no adequate reason why it should not be more generally employed, for although its application to triangular surfaces necessarily gives double areas,—a division by two is the briefest imaginable.

Right and left of centre each triangle is obviously equal to *half* the rectangle of the hight and width on that side (the triangle and rectangle having a common base, and lying between the same parallels, *a* and *b*), and by addition, *the double area of the whole trapezium =* *hight* $\times$ *width.*

(**b.**) ...... In view of the rule just recited, for finding the areas of triangles and trapeziums, by hights and widths, it becomes of some importance to have a concise rule* for determining the *distances out* of the vertices from the axis, when the hight and slopes alone are

---

* Gillespie, Roads and Railroads (1847), gives rules analogous to ours, but they had long before *been known.*

given: in this there is little difficulty, as engineers have long been possessed of formulas for the purpose, *similar* to those which will be seen below, referring to *Figs.* 12 and 13,—*and these distances out*, when added together, form the width *w*, of the rule above.

In *Fig.* 12.

$$\frac{\overset{\text{Ht.}}{40} \times \overset{\text{Wid.}}{60\cdot 8}}{2} = \frac{\overset{\text{Area.}}{2432}}{2} = 1216.$$

Both in trapeziums and triangles the diagonal $\times$ the sum of perpendiculars from the opposite angles $=$ *double area*.

Or, centre hight $\times$ the total width $=$ *double area*.

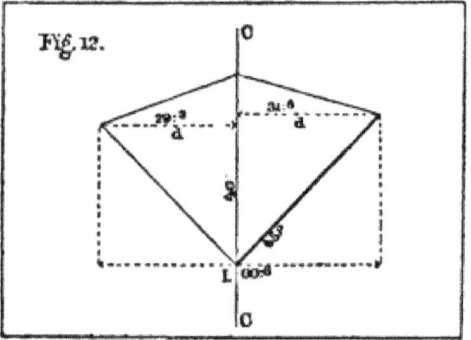

Suppose, in both these figures, the side-slopes, ground-slopes, and centre hight, or axis, *given*, and the side-slopes intersected at I, then to find the *distances out*, right and left of centre, *take each side separately.* Consider the centre line, or axis, to be a meridian (*as in a map*), imagine also an east or west line, drawn through the origin of each slope (*side or ground*).

Then,

If the slopes incline towards the *same* compass quarter:

$$\frac{\text{Hight}}{\text{By difference of nat. tans. of slopes}} = distance\ out = \mathbf{d.}$$

If the slopes incline towards *adjacent* compass quarters:

$$\frac{\text{Hight}}{\text{By sum of nat. tans. of slopes}} = distance\ out = \mathbf{d.}$$

These results on both sides of centre, added together, give the total width *of the whole trapezium.*

In *Fig.* 13.

$$\frac{\overset{\text{Ht.}}{30} \times \overset{\text{Wdt.}}{88\cdot2}}{2} = \frac{\overset{\text{Area.}}{2646}}{2} = 1323.$$

These rules also furnish a concise and easy method of finding *the half breadths*, a matter deemed quite important by foreign engineers.

(c.) ...... The side slopes (bounding the diedral angle) remaining plane surfaces as usual in the cross-sections of earthwork, we sometimes find the ground surface *very irregular*, but even these cases, upon the principle of *equivalency*, may be correctly dealt with, so as to reduce them easily to the plane figures of the elements of geometry.

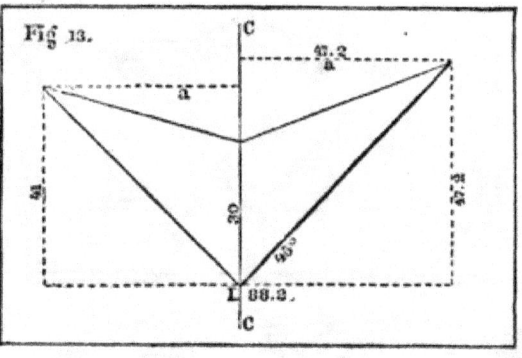

Thus, although, as far as we have shown, the rule of $\frac{h\,w}{2}$, applies only to a line *once broken*, so as to change the figure considered, from an oblique triangle into a trapezium ; nevertheless, it is not difficult to reduce or equalize a surface line, *very much broken*, by a single one properly drawn, which shall contain within it an area *exactly equal* to that bounded by the irregular outline, and thus bring it within the rule.

In *Fig.* 14, let ABCDEFGH be the cross-section of a railroad cut, base 20, slopes 1 to 1, intersecting at I, the centre line being marked CC—(this area looks irregular enough, but had it been ten times more so, the process below *would have equalized it exactly*.)

Then, from the top of the shortest side hight at H (adopted for convenience), draw a line HK parallel to the road-bed, or base AB, making a level trapezoid 10 feet high upon the section, or ABKII = 300 *in area.*

Now, we will find, by a common calculation, the area of the whole cross-section—between base AB, side slopes, and broken ground line —*to contain* = 654 *area.* Neglecting in this case the grade triangle at I, as being a common quantity, not affecting the result :—(but adding the grade triangle (100), the area, from the ground line down to the edge of the diedral angle at I = 754).

Then, 654 — 300 = 354, the area of the partial cross-section above IIK, extending to the irregular outline, which is to be *correctly equalized,* by a single sloping line drawn from II.

Now, $\dfrac{354}{\frac{1}{2}\,\text{IIK}} = 17\cdot 7 = \text{LM}$, the altitude of a triangle IIKM, on the base IIK, which is *exactly equivalent* in area to the partial cross-section above HK.

So that IIM is a single equalizing line, drawn from II, equivalent to the broken line of ground, and including the same area *exactly.* Another way of finding the point M — the terminus of the equalizing line—is the following : $\left\{ \dfrac{\text{Double area} = 1508}{\text{III} \times \sin. \text{ of I}} = 53\cdot 3 \right\} \dots \dots$ and this is a very concise method, as III is easily found.*

***

* This rule will be found useful as a *verification* of the process of *Fig.* 14.

If the *degree* of equivalent surface slope be desired (*as it usually is*),

Then, $\dfrac{57\cdot7}{17\cdot7} =$ cot. 17° (nearly) = 3·26.

The slope of the equalizing line HM being 17° ascending from H, we easily find FN =6·135, and adding FI = 20, we have IN or $h =$ 26·135, and $w = 57\cdot7$. Then, $\dfrac{h \times w = 26\cdot135 \times 57\cdot7}{2} = 754$, and

deducting the grade triangle (ABI = 100), we have, finally, the area of the whole cross-section above the road-bed = 654, thus verifying

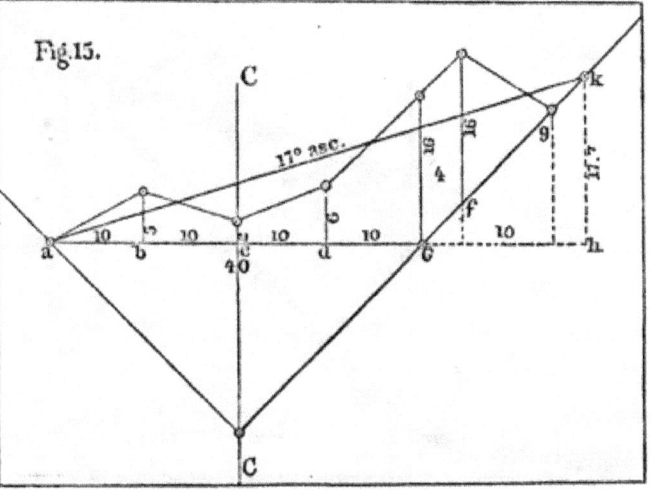

Fig.15.

the original calculation as before given, and, by using the radii of inscribed and circumscribed circles, we can prove it, *if necessary:* (*Fig.* 14).

(**d.**) . . . . . . It is sometimes desirable, by means of an equalizing line, to deal with the boundary *alone*, without the rest of the cross-section, and this is not difficult, for we may consider the broken line HKM (*Fig.* 14), or *a e g* (*Fig.* 15), as a base of ordinates, preserving, however, their parallelism, and taking all the distances horizontally as though the base were straight (see *Fig.* 15); but the process of *Fig.* 14 is generally preferable.

It is often useful to equalize a section by a level top line, *or slope of* 0°. This can be done as shown in *Art.* **6.**

Whole area . . . . $=$ *a.*

Slope ratio . . . . $=$ *r.*

Level hight . . . . $=$ *h.*

Then h . . . . $= \sqrt{\dfrac{a}{r}}.$

The ordinates marked upon *Fig.* 15 are deduced from those of *Fig.* 14, and the calculations of the irregular area, $a\,e\,g$, are made by successive trapezoids, and double areas, as follows:

Ordinates in pairs above the base line, *a e g*, broken at *e*	$a+b$ $0+5$ 5	$b+c$ $5+2$ 7	$c+d$ $2+6$ 8	$d+e$ $6+16$ 22	$e+f$ $16+16$ 32	$f+g$ $16+0$ 16
Horizontal distances apart . . . . . . $=$	10	10	10	10	4	10
Double areas (total 708) . . . . $=$	$\overline{50}\ +$	$\overline{70}\ +$	$\overline{80}\ +$	$\overline{220}\ +$	$\overline{128}\ +$	$\overline{160}$

Then,*

$$\frac{\text{Sum of double areas} = 708}{\text{Base of equalizing triangle, } a\,e = 40} = 17{\cdot}7 = h\,k,\ as\ before.$$

And $a\,k$ is the equalizing line, ascending from a, with a slope of 17°, which is equivalent to HM, of *Fig.* 14.

(e.) . . . . . We may now briefly refer to the computation of cross-

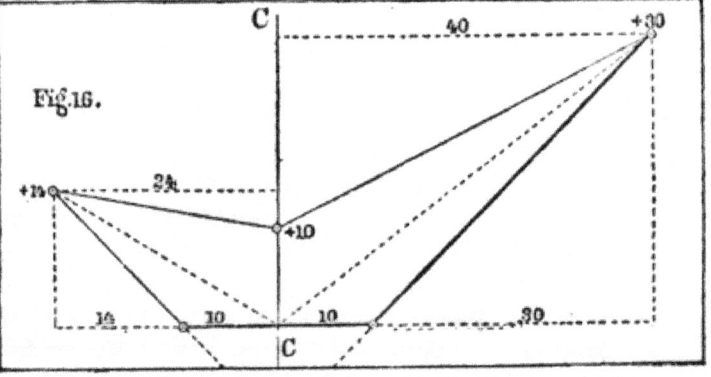

Fig. 16.

sections. These are usually taken in the field with the *rod, level, and tape;* they designate by levels, and distances out, the prominent

---

* With *equal* abscissæ, Simpson's well-known rule, or that of Davies Legendre, would conveniently apply.

points, or features of the ground, and fix the intersection of the side slopes, or place of the slope stake, which bounds the limits of excavation or embankment; and on regular ground, the clinometer may be used, *but is less correct and satisfactory.*

On plain ground, but *three* levels are taken,—the centre and side hights,—and this has been called *three-level ground.* It is the practice of many engineers (and it is a good one) to take angle levels and distances over the edges of the road-bed, this then becomes *five-level ground;* and where more than five levels are necessarily taken, the cross-section is usually deemed *irregular,* though the point where sections become irregular is not well defined, and may be safely left to the judgment of the engineer.

In this case (*Fig.* 16), the centre and side hights, and the right and left distances out to the slope stakes, are always given, and the calculation becomes simple and rapid.

The following is the method long ago used by engineers, and published by Trautwine * and others, twenty years since.

RULE for area of cross-section, with uniform road-bed and centre and side hights given.

Half the centre cutting $\times$ by right and left distance, *plus* right and left cuttings $\times$ one-fourth of road-bed.

Thus, in *Fig.* 16,
We have, by this rule,

$$5 \times 64 = 320.$$
$$44 \times 5 = 220.$$
$$Area.. = \overline{540.}$$

And by using the grade triangle and hights and widths, as in *Figs.* 10 and 11, *We have,*

$$h = 20.$$
$$w = 64.$$
$$\frac{hw}{2} = \frac{20 \times 64}{2} .. = 640.$$
Less grade triangle . $= \overline{100.}$
$$Area. .... = \overline{540.}$$

(f.) ...... To find the area of cross-sections, where angle levels have been taken,† or *five-level ground* (which angle levels have long been used by engineers, and are recommended by Prof. Davies in his new surveying), we will give an example for illustration, from which the rule of this method will be evident. (See Cross, Eng. Field Book, N. Y., 1855.)

---

* Trautwine's New Method of Ex. and Em. (1851).

† Davies' New Surveying (1870),—cross-section levelling.

Now, to calculate the area of this cross-section, *Fig.* 17, by double areas,

We have,		*Equivalent to,*

By dividing the figure into six triangles, or *three trapeziums.*

$$\left\{ \begin{array}{r} 20 \times 15 = 300. \\ 20 \times 12 = 240. \\ 34 \times 16 = 544. \\ \hline 2)\overline{1084.} \\ Area. = 542. \end{array} \right\}$$

Triangle, 15×10 . $= 150.$
Trapezoid, 27×10 . $= 270.$
" 28×10 . $= 280.$
Triangle, 16×24 . $= 384.$
$2)\overline{1084.}$
Area. . . . . $= 542.$

To compute this area in the usual method by successive trapezoids and deductive triangles, *is much longer and less satisfactory.*

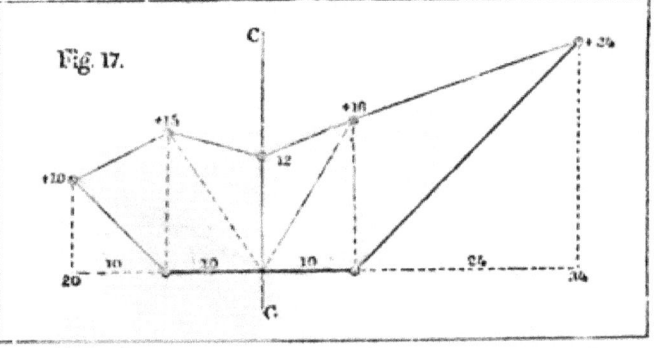

Fig. 17.

(g.) . . . . . . For *very irregular* cross-sections, no definite rule can be given,—they are usually reduced to elementary forms, which, being separately computed, and finally totalized, give the whole area in the end.

This reduction is usually made to trapezoids and triangles (*additive or deductive*), while the calculations are the simplest possible, though, from the multitude of figures, *necessarily tedious.*

In the most irregular sections, involving heavy rock-work on side-hill,—the several cuttings (or level hights), transversely, are frequently taken at ten feet only, or some such *uniform* distance apart, and in these cases the mean hights of a number of contiguous trapezoids may be ascertained, and multiplied by the *uniform* distance (agreeably to the rules of mensuration for irregular areas), and thus abbreviate somewhat the labor of such computations; which, however, in their origin, and indispensable verifications, *are often laborious enough,* though, fortunately, so simple and elementary as to be within the comprehension of *all* the members of an engineer party, which enables us to bring many hands to the work.

Not unfrequently, too, in rock-work (proximating a cost of a dollar per cubic yard), it has been deemed necessary to take independent cross-sections, at only *ten feet apart forward*, over the roughest portions of the work.

In that event, although the calculations become voluminous, we have the satisfaction of knowing *that the solidity is correctly obtained;* since, in such short spaces, no ordinary rules would produce any important variation in the final result; supposing, of course, the cross-sections to be correctly laid out, and measured with accuracy, both horizontally and vertically—a matter of no small difficulty on steep, rocky hill-sides, *when cleared for work.*

9. *Further Illustration of the Modification of Simpson's Rule—*(**II.**), *with a Diagram Representing it, and also one of the Regular Formula, and another Modification.*

Here let us take the triangular prismoid, *cross-sectioned,* in *Fig.* 8 (and shown below), and suppose its length 100 feet (h)—the end

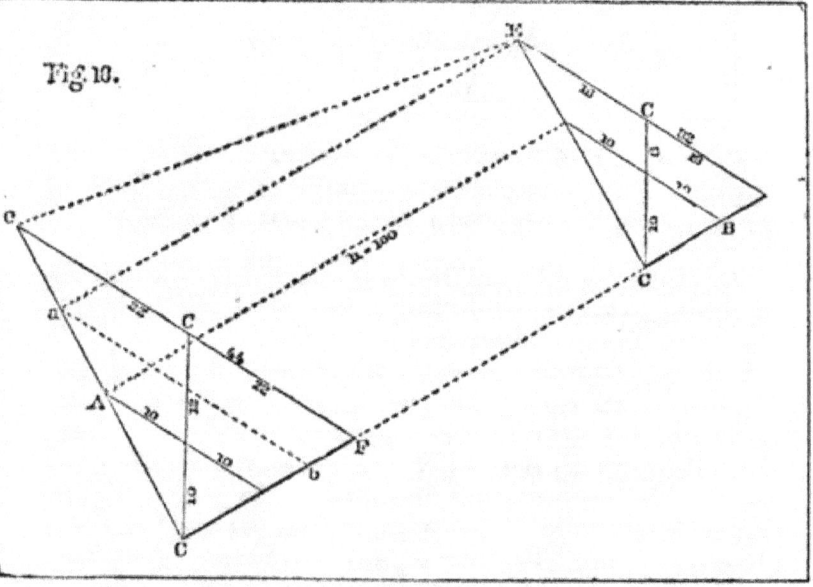

Fig. 10.

cross-sections being dimensioned as before. With road-bed of 20, and slopes of 1 to 1. The whole, shown in projection, to give a better idea of the nature of *the solid.*

References.

CC   = Centre line and edge diedral angle.
ACCB = Grade prism.
AB   = Road-bed, 20.
AE   = Side-slope plane, 1 to 1.
EF   = Ground plane, assumed as level.
$e\,a\,b\,E$ = Wedge of *Fig.* 8.

Then, for the volume of this solid, we have, by the modification of Simpson's Rule (**II.**),

$$\left\{\begin{array}{l}
\text{Hights. Widths.}\\
\text{Near end (double area), } 22 \times 44 \quad . \quad . \quad . = \quad 968 = 2\,b.\\
\text{Far end,} \qquad\quad `` \qquad\quad 16 \times 32 \quad . \quad . \quad . = \quad 512 = 2\,t.\\
\text{8 times mid-section, . . } 38 \times 76 \\
\qquad\qquad = \text{ sum hts.} \times \text{ sum wids.}\left.\begin{array}{l}\\ \end{array}\right\} = 2888 = 8\,m.\\
\qquad\qquad\qquad\qquad\qquad\qquad 12)4368\\
\qquad\qquad\qquad\text{Mean area. . . } = \quad 364\\
\qquad\qquad\qquad\quad\text{Length } h. \;. \;. = \quad 100\\
\text{Whole triangular solid to intersection}\left.\begin{array}{l}\\ \end{array}\right\} = \quad 36400\\
\quad\text{of slopes. } . \;. \;. \;. \;. \;. \;. \;. \;.\\
\text{Deduct grade prism } under \text{ road-bed. . } = \quad 10000\\
\text{Leaves volume } above \text{ road-bed, or } Trape\text{-}\left.\begin{array}{l}\\ \end{array}\right\} = \quad 26400 = The \; same\\
\quad zoidal \; Prismoid \; of \; Earthwork. \;. \;.
\end{array}\right.$$

solidity, as before computed, Art. **e.**

(**a.**) . . . . . . The transformation or modification of Simpson's Rule (**II.**) *may, in its mid-section term,* be conveniently represented by a diagram (perhaps more curious than useful).—*Thus,* continuing the side-slopes through the intersection, so as to form the end cross-sections, *one above the other.*

So, in *Fig.* 19, dimensioned as in *Fig.* 8, we have,

The triangle IEF   = The larger end section, or area.
  "   "   ICD   = The smaller one.
  " rectangle KLMN = 8 times the area of the mid-section,
                      or the circumscribing rectangle
                      formed by *sum of hights* × *sum
                      of widths.*
The road-beds . . . = The dotted lines, and may be
                      assumed (parallel) anywhere.

The parallelogram IFEP = Hight × width of larger end, or *double area* of . **A.**

" " IDCO = Hight × width of smaller, or *double area* of. . . **B.**

" rectangle   KLMN = HG × OP, or sum hights × sum widths, = 8 times the mid-section.

Here it is evident that HI × FE = Double area of larger end section, or = IFEP . . . . . . and IG × CD = same of smaller = IDCO.

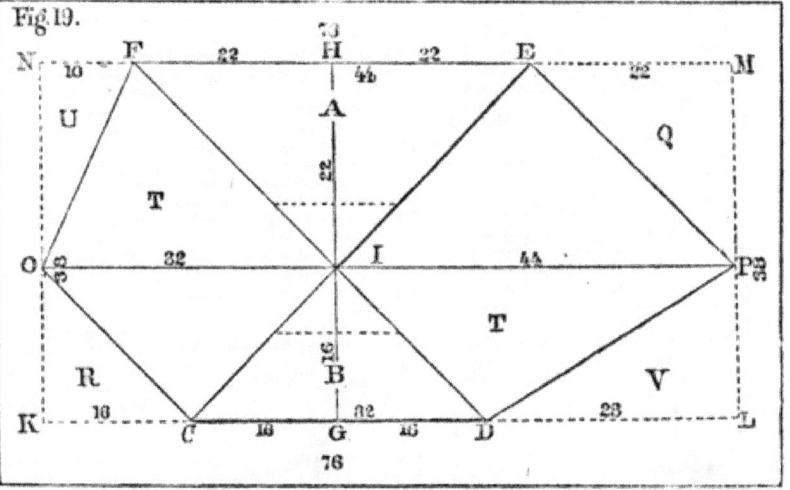

Fig.19.

While (CD + FE) × (GI + IH) = the circumscribing rectangle KLMN = HG × OP, or the rectangle of sum of hights and sum of widths.

Also,

$$\begin{cases} \left(\dfrac{HI + IG}{2}\right) \times \left(\dfrac{FE + CD}{2}\right), \text{ or } \dfrac{19 \times 38}{2} = 361, \text{ the mid-sec.} \\ HG \times OP, \text{ or } 38 \times 76 \quad . \quad . \quad . \quad . \quad = 2888, \text{ or 8 times mid-sec.} \end{cases}$$

The triangles Q and R taken together = *the Arithmetical Mean* of A and B, the end areas = (16 × 8) + (22 × 11) = 128 + 242 = 370, or

$$\frac{484 + 256}{2} = \frac{740}{2} = 370, \text{ *the Arithmetical Mean.*}$$

The triangles T and T are *each* equal to the *Geometrical Mean* of the end sections A and B = $\sqrt{484 \times 256} = 352$.

While U and V added together proximately equal the *Harmonic Mean* between A and B, or = 334.

So that the circumscribing rectangle, KLMN, representing the mid-section term, of Simpson's Transformed Rule (**II.**), *contains, or is composed of, the following areas.*

$$
\left\{
\begin{array}{l}
\left\{
\begin{array}{l}
\text{Double area of A.} \quad . \quad . \quad . \quad . \left\{ \begin{array}{l} 484 \\ 484 \end{array} \right. \\
\text{``} \quad \text{``} \quad \text{B.} \quad . \quad . \quad . \quad . \left\{ \begin{array}{l} 256 \\ 256 \end{array} \right. \\
\text{(The two end sections.)}
\end{array}
\right. \\
\text{Arithmetical Mean.} \quad . \quad . \quad . \quad . \quad . \quad 370 \\
\text{Geometrical Mean} \times 2. \quad . \quad . \quad . \left\{ \begin{array}{l} 352 \\ 352 \end{array} \right. \\
\text{Harmonic Mean.} \quad . \quad . \quad . \quad . \quad . \quad 334 \\
\qquad \text{\textit{Total 8 times the mid-sec.,}} \\
\qquad \qquad \text{\textit{or} } 361 \times 8. \quad . \quad . \quad = 2888
\end{array}
\right.
$$

In this case :
= Double areas
of both ends +
4 times the Geo-
metrical Mean
= 2888.

Some curious inferences may be drawn from this diagram, but their practical results can be more concisely obtained in other forms.

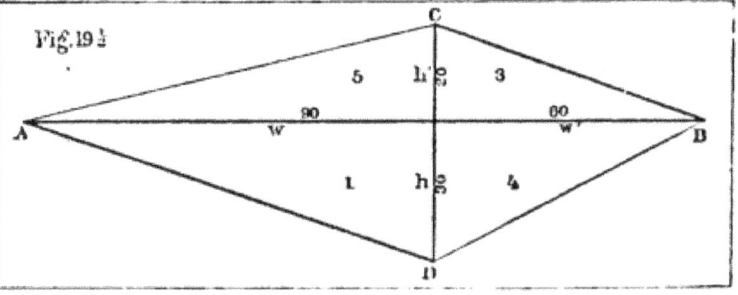

Diagram of the regular Prismoidal Formula of Simpson and Hutton.

As applied to a triangular prismoid, formed by a diagonal cutting plane, from the rectangular prismoid, *Fig.* 2, and shown again in *Figs.* 22, 24, and 52, with side-slopes of 1½ to 1.

Let 1 (*Fig.* 19½) Be the larger end section (*Fig.* 22), transformed
     into an equivalent right triangle.

     3 The smaller end (*Fig.* 24), also transformed:—4 and 5,
     additive triangles, making up the trapezium ABCD (*Fig.*
     19½), equivalent in area *to four times the prismoidal mid-*
     section (*Fig.* 23).

From this diagram we readily deduce a simple modification of the
prismoidal formula, equivalent in result, for triangular prismoids.

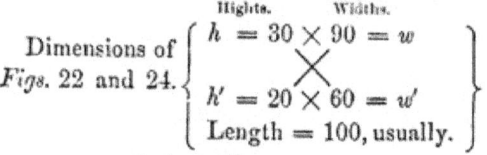

$$\text{Dimensions of } \textit{Figs.} \text{ 22 and 24.} \begin{cases} h = 30 \times 90 = w \\ \qquad \times \\ h' = 20 \times 60 = w' \\ \text{Length} = 100, \text{usually.} \end{cases}$$

Then, $\quad \dfrac{hw + hw' + \left(\dfrac{hw' + h'w}{2}\right)}{6} \times \text{length} = Solidity. \quad$ **VIII.**

This operates very simply in figures, by *direct and cross multiplica-*
tion of hights and widths.

Substituting the numbers, *Solidity* = 95000, as hereafter computed,
Art. **10** (a).

10. *Adaptation of the Prismoidal Formula to the Quadrature and*
Cubature of Curves, and also Solids, where the Ordinates are equivalent
to Sections—by the Method of Simpson, as explained by Hutton.

The eminent mathematician, THOMAS SIMPSON, to whom we are
indebted for *the Prismoidal Formula,* also devised a method for the
quadrature of irregular curves by means of equidistant ordinates, or
for their *cubature,* by using equivalent sections of irregular solids, *at*
equal distances, instead of ordinates; such solids being bounded opposite the base by a general curved outline.

This method, although a century old, is still the simplest and best
yet known for proximating the area of irregular curves, or the volume
of unusual solids,—it has attained great celebrity, and been of much
service to philosophers and calculators, ever since its origin in 1750.

It has long been used by military engineers for ascertaining the
volume of warlike earthworks, and is regularly quoted in the leading
text books of that important profession.*

Also by naval architects in determining the nice problem of the
displacement of ships; by mechanical philosophers, like Morin and

---

* Laisné, Aide Mémoire, du Génie.—Eds., 1831-61.

Poncelet, etc.—by these it has been deemed of much importance, not only for the quadrature of irregular areas, but also for the "Cubature of solids of irregular excavations, embankments, etc." *

It forms a leading feature in Hutton's remarkable chapter on the cubature of curves (who seems to have fully adopted it), under the name of *the method of equidistant ordinates.*—(See 4to Mens., 1770, sec. 2, part iv. page 458.)—We are much indebted to Hutton for the practical development of this important problem, and he gives several examples of its utility. Amongst others, computing the area of a quadrant of a circle, with radius $= 1$,—which, by Simpson's method, using 11 ordinates, gives ·7817 area, instead of ·7854—"*pretty near the truth*" (says Hutton).

We will describe this method from the—(4to Mens., 1770, p. 458).
"If any right line, AN, be divided into any even number of equal parts, AC, CE, EG, etc., and at the points of division be erected perpendicular ordinates, AB, CD, EF, etc., terminated by any curve, BDF, etc."

Then, the sum of the first and last ordinates, *plus* 4 times sum of even ordinates, *plus* 2 times sum of odd ones, $\div$ by 3, and $\times$ by AC, *one* of the equal parts; the resulting product will equal the area, ABON, "*very nearly.*"

That is to say, if

$$
\left\{
\begin{array}{l}
\text{The sum of the } \textit{two} \text{ extreme ordinates . . } = \text{A.} \\
\quad\text{"} \quad \text{of all the } \textit{even} \text{ numbered } \quad \text{"} \quad . = \text{B.} \\
\quad\text{"} \quad \text{of all the } \textit{odd} \text{ numbered } \quad \text{"} \quad . = \text{C.} \\
\text{The } \textit{common distance} \text{ apart of ordinates . . } = \text{D.}
\end{array}
\right\}
\begin{array}{l}
\text{(Excepting} \\
\text{the first and} \\
\text{last from C.)}
\end{array}
$$

Then the rule is,

$$\frac{A + 4B + 2C}{3} \times D \text{ (or AC)} = \text{Area, ABON.} \quad . \quad . \quad . \quad \textbf{(IX.)}$$

And if more convenient (*as it may be*), we transform this *into its equivalent,*

$$\frac{A + 4B + 2C}{6} \times 2D \text{ (or AE)} = \text{Area, ABON.} \quad . \quad . \quad \textbf{(X.)}$$

n applying this formula, it is desirable to draw a figure, and number all the ordinates (as below), commencing with 1.

---

* Morin's Mechanics (Bennett's Trans., 1860).—See also Gregory, Math. Prac. Men. (1825).

"The same theorem will also obtain, for the contents of all solids, by using the sections perpendicular to the axe, instead of the ordinates."

In this form it becomes applicable to excavations and embankments, or any similar solids relating to a guiding line, centre, or base line, to which the cross-sections representing ordinates are perpendicular.

See *Fig.* 20, copied below from Hutton, page 458.

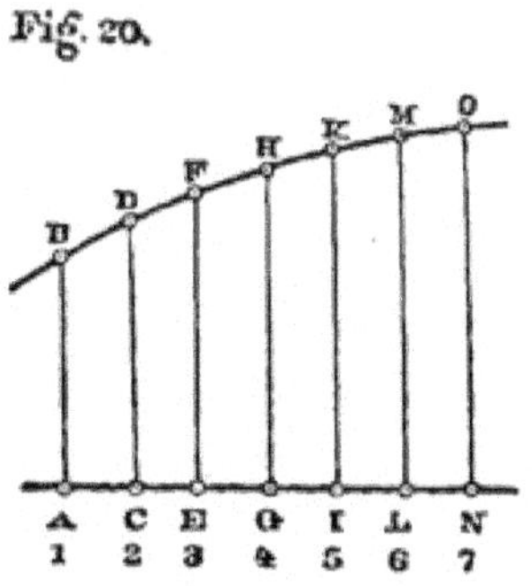

Hutton's Example 3, p. 462.

"Given the length of five equidistant ordinates of an area, or sections of a solid, 10, 11, 14, 16, 16, and the length of the whole base, 20."

Then,

$$\frac{26 + 108 + 28}{3} \times 5 = 270.$$

"*The area or solidity required.*"

Fig. 20.

This formula of Simpson (adopted by Hutton) is evidently derived from *the Prismoidal Formula*, or it may be, *originated it*, both having the same author, and their precedence unknown.

(a.) . . . . . . We will now give an example of Hutton's *Method of Equidistant Ordinates* (adopted from Simpson),—giving two stations of a railroad cut (each 100 feet long, with a road-bed of 18, and side-

Fig. 21.

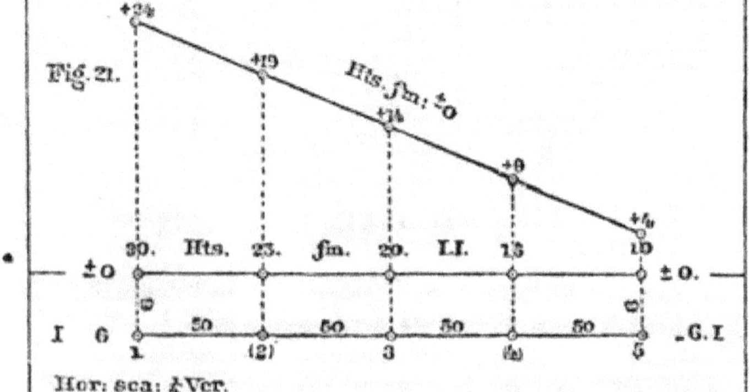

slopes $1\frac{1}{2}$ to 1), shown both in profile and cross-sections. (See *Figs.* 21 to 26, inclusive.)

The above figure is a profile, or vertical section (of two stations), upon the centre line of a railroad cut, with a road-bed of 18, and side-slopes of $1\frac{1}{2}$ to 1. The horizontal scale (*for convenience*) being made $\frac{1}{4}$ of the vertical.

Firstly : Computing each station separately, by Simpson's Rule (**II.**)

Stations 1 to 3 $=$ 100 $= h.$ | Stations 3 to 5 $=$ 100 $= h.$

Hts.	Wids.				Hts.	Wids.		
30 $\times$	90 $=$	2700	$= 2\,b.$		20 $\times$	60 $=$	1200	$= 2\,b.$
20 $\times$	60 $=$	1200	$= 2\,t.$		10 $\times$	30 $=$	300	$= 2\,t.$
50 $\times$	150 $=$	7500	$= 8\,m.$		30 $\times$	90 $=$	2700	$= 8\,m.$

$\div$ by 12)11400 $\div$ by 12)4200

Mean Area . . $=$	950	Mean Area . . $=$	350
$\times$ by h . $=$	100	$\times$ by h . $=$	100
Solidity in c. ft. $=$	95000	Solidity in c. ft. $=$	35000
$\div$ 27 . . $=$	3519	$\div$ 27 . . $=$	1296
Deduct Grade		Deduct Grade	
Prism for 100		Prism for 100	
feet . . . $=$	200	feet . . . $=$	200
Solidity in c. yds. $=$	3319	*Solidity in c. yds.* $=$	1096

Then, $3319 + 1096 = 4415$ *cubic yards*, whole solidity of cut from 1 to 5 inclusive.

Secondly: Now computing the same, *in a body*, by Hutton's Rule (**X.**).

Data.

$$
\left\{
\begin{array}{l}
A = \left\{ \begin{array}{l} 1350 \\ 150 \\ \hline 1500 \end{array} \right. \\[3em]
B = \left\{ \begin{array}{l} 937\cdot5 \\ 337\cdot5 \\ \hline 1275 \ \times\ 4\ =\ 5100 \end{array} \right. \\
C = \phantom{\{} 600 \ \times\ 2\ =\ 1200
\end{array}
\right\}
$$

We have, $\dfrac{1500 + 5100 + 1200}{6} \times 100 =$ C. feet. $130{,}000$

Now, $\div$ by 27 . . . . $=$ 4,815

Deduct Grade Prism, 200×2 stations. $=$ 400

Solidity in cubic yards . . . . . $=$ 4,415

(The same as above.)

(CROSS SECTIONS.)

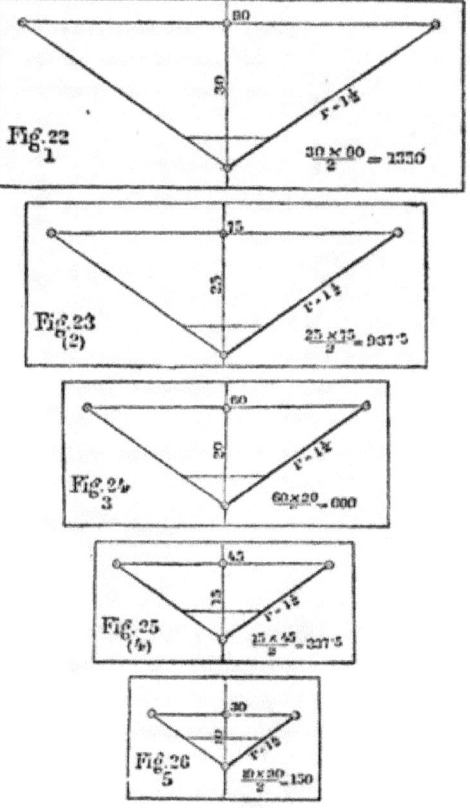

(b.) . . . . . . The preceding example clearly shows that Hutton's method of equidistant ordinates is merely the Prismoidal Formula extended to several stations, *instead of confining it to one.*

There is another mode of considering this question where the cross-sections are *triangular,* and the ground *level transversely.*

Thus, in any station, let h and h' be the end hights from the intersection of the side-slopes to the ground, then, $h^2 r$ and $h'^2 r =$ the corresponding areas (r being the slope ratio, which, in the preceding example $= 1\frac{1}{2}$), then omitting r, *a common factor,* we have in h^2 and h'^2 vertical lines, or ordinates, *representative* of the end areas, and in $\left(\dfrac{h + h'}{2}\right)^2$ of the mid-section.

The square roots, then, of the areas (however computed, and whatever be the ratio (r) of the side slopes), correctly represent them; since these roots form the side of an equivalent square (or half base of an equivalent triangle, with 1 to 1 side-slopes)—*squaring which*, obviously re-produces the areas they are the roots of.

Hence, the end areas being given in any station, or number of stations, their square roots may represent them in Hutton's rule of cubature, and any pair of roots added together, *and their sum squared*, gives 4 times the mid-section between them; which is precisely what we need in *the Prismoidal Formula.*

This is evident, from *Fig.* 27, where we suppose h and h' placed in a continuous line, then, $\left(\dfrac{h + h'}{2}\right)^2 = \frac{1}{4}$ the square of (h

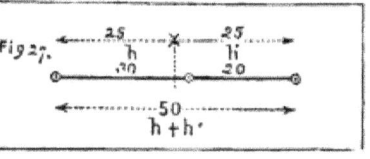

+ h'), or equivalent to the proposition of geometry—*that the square of a whole line equals 4 times the square of half.*

$$\left\{ \begin{array}{l} \text{Let } h = 30, \text{ and } h' = 20, \text{ then } h + h' = 50, \dfrac{h + h'}{2} = 25 \\[2mm] \left(\dfrac{h + h'}{2}\right)^2 = (25)^2 = \text{ the mid·sec.} = 625, \text{ and } \times 4 = 2500 \\[2mm] (h + h')^2 = (50)^2 \quad .\;.\;.\;.\;.\;.\;.\;.\;.\;.\; = 2500 \\[2mm] \text{While } h^2 = 900 = \text{ one end area, and } h'^2 = 400, \text{ the other.} \end{array} \right\}$$

Also,

$$\left\{ \begin{array}{l} h^2 + h'^2 + 2\,(h \times h') \\ = 900 + 400 + \quad 1200 = 2500 \\ = (h + h')^2 \quad .\;.\;.\;.\; = 2500 \end{array} \right\}$$

From all which, we readily draw the following:

Rule.—Compute the end areas at each *regular station* (numbered upon a diagram on Hutton's plan, by the *odd* numbers, 1, 3, 5, 7, etc., marking also the *even* numbers *intermediately*, which are, in fact, half stations, or the places of mid-sections),—find the square roots of these end areas—add any two adjacent roots, and *their sum squared* equals 4 times the area of the *mid-section*, between the regular stations.

Let *Fig.* 28 be the profile of *one station of cutting*, from intersection
of slope to ground.

h and h' = The end hights, or representative square roots of
the areas, at regular stations, numbered *odd*.

m = The place of the mid-section, numbered *even*, and represented by *its ordinate*.

Length = usually, 100, between principal stations.

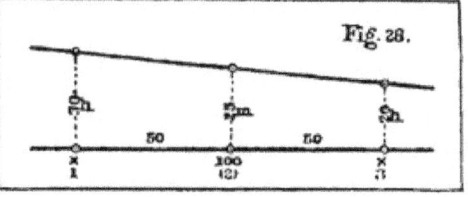

Fig. 28.

Whence,

$$\begin{cases} \dfrac{h^2 + h'^2 + 4\,m^2}{6} \overset{\text{Length.}}{\times} 100 = \textit{Solidity}, \text{ by } \textit{the Prismoidal Formula.} \\[2mm] \text{Or,} \ \dfrac{h^2 + h'^2 + (h + h')^2}{6} \overset{\text{Length.}}{\times} 100 = \textit{Solidity.} \ \ \ldots \ldots \ \mathbf{XI.} \end{cases}$$

Which, for one station, is equivalent to Hutton's Rule.

(**c.**) . . . . . . So that having the end areas given, we deduce at once
the mid-section, by a table of roots and squares,* and can proceed
station by station, *prismoidally*, to find *the solidity.—Or combining
them as in Hutton's Rule for cubature, we may calculate in a body the
whole of a cut or bank.*

Thus, taking the preceding example, and tabulating it (see *Figs.*
21 to 26).

Stations.		Areas.		Roots.	Sums.	Even Nos.
Odd.	Even.	Extreme.	Odd Nos.			Squares, or Mid-sec. Areas.
1		1350		36·7423		
	2		600		61·24	3750
3				24·4949		
	4				36·74	1350
5		150		12·2475		
		1500	600			5100
			2			
			1200			
		A.	2 C.			4 B.

This tabulation may be made in any more convenient form, or the
data may be written upon the working profile of the line with advantage.

---

* Such as Barlow's (Prof. De Morgan's Ed., London, 1860), which is the most convenient and extensive,—*or any like tables.*

Then,

$$\left\{ \begin{array}{l} \text{A} + 4\,\text{B} + 2\,\text{C} \quad {\scriptstyle\text{Mean Area.}\ \ \text{Length of Sta.}\ \ \text{Cub. Ft.}} \\ \dfrac{1500 + 5100 + 1200}{6} = \dfrac{1300 \times 100 = 130000}{\textit{Rule }\mathbf{X}.} = \textit{by Hutton's} \end{array} \right.$$

Now, dividing by 27, ... =   4815

Deduct grade prism for two stations . . =   400

Leaves *solidity* in cubic yards (as before) =   4415.   From 1 to 5 = 200 feet.

The division by 6 in the first term results in *a mean area*, which $\times$ by length, gives *the solidity*—and enables us to use a table of cubic yards to mean areas, as soon as we have found the latter, in order to obtain the cubic yards more readily *by inspection*.

(d.) . . . . . . In further illustration of this important method of computation in earthworks,—we will submit another example, representing an entire railroad cut, with 20 feet road-bed, and side-slopes of 1 to 1, laid off in regular stations of 100 feet, and truncated at both ends in light cutting (at selected stations), so as to secure full cross-sections *throughout;* and also an even number of equal distances (apart sections), each 100 feet, or regular and uniform stations, whatever their length.

These truncations are made before proceeding to the calculation, so that all the cross-sections shall be *complete* (or have some side slope—*however small*—at both edges of the road-bed), which simplifies the main calculation, while in the end the truncated volumes may be computed independently, and added in with the rest.

Again, if the ground should have required the insertion of *intermediates* in any one or more of the regular stations, it will be best to draw a pencil line around all such whole stations upon the diagram, and compute them separately from the main body—the places of such stations being considered vacant for the time (omitting distance, midsection, and end areas, so far as they apply to the assumed vacancy), and thus the cut will be computable under our rule, in one or more masses (as though a single mass originally), according to the number of vacant spaces. A little practice will familiarize this matter better than further explanation, *as the object to be attained is evident.*

Generally, we may compute the cut, or bank, in one principal mass, and then calculate separately, *and add.*

1. The solidity in the special stations containing intermediates.

2. The quantities of work of the same kind, at the passages from excavation to embankment, at both ends of the cut (as will be further explained).

In all such cases (*indeed, in all cases of heavy work*), it is necessary to draw diagrams, as below, and these (in cross-sections) will usually have a scale of 20 feet to the inch, which long practice has shown to be entirely suitable; but any preferred scale may be employed, or the cross-section paper in common use amongst engineers—which carries its own scale—and which will be found convenient in many respects, either bound up for the purpose, or in loose sheets, to be ultimately tacked together, including a mile forward, or thereabouts.

Profile of 8 stations of railroad cut; base 20, side-slopes 1 to 1.

$\Big\{$ $a\,b$ = Intersection of side-slopes, or edge of diedral angle, formed by their planes meeting.

$c\,d$ = Grade, or formation line of the road-bed = $\pm$ 0·0.

$e\,f$ = Surface line of ground, as cut by centre plane.

$g\,p$ = Grade prism—*deductive for solidity*

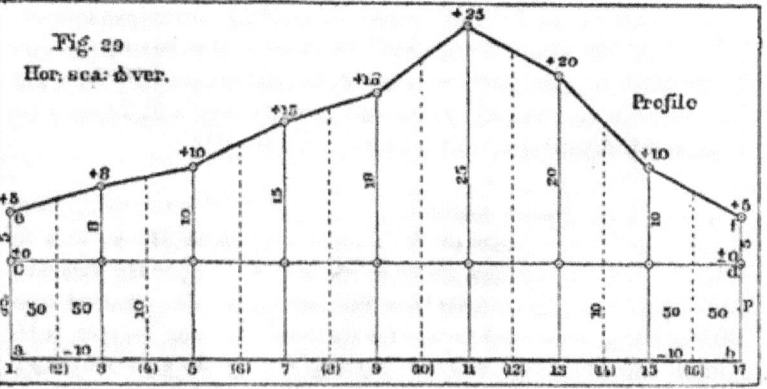

Regular stations designated by *odd* numbers (1, 3, 5, etc.).

Mid-section places by *even* numbers (2, 4, 6, etc.)

The ordinates show *the level hights* from grade to ground, to which *add* always the *common hight of grade triangle.*

Transverse slopes are shown on cross-sections.

Regular Stations	=	1·		3·		5·		7·		9·		11·		13·		15·		17·
Cross-section Areas	=	232·5		349·2		412·7		720·5		844·8		1085·		901·5		516·		259·5
Square Roots	=	15·25		18·60		20·31		26·84		29·06		32·94		30·02		22·72		16·09
Sums of Roots	=		33·94		39·00		47·15		55·90		62·00		62·96		52·74		38·81	
Squares of Sums	=		1151·9		1521·0		2223·1		3124·8		3844·0		3964·0		2781·5		1506·2	

These squares are each equal to 4 times the mid-section, between regular stations.

All hights and areas taken to intersection of slopes.

Mean areas computed separately for each regular station, by Simpson's Rule.

$$
\begin{array}{r}
232·5 \\
(1 \text{ to } 3) \quad 349·2 \\
1151·9 \\
\hline
6)1733·6 \\
\hline
\text{Mean Area} = \quad 288·9
\end{array}
$$

$$
\begin{array}{r}
349·2 \\
(3 \text{ to } 5) \quad 412·7 \\
1521·0 \\
\hline
6)2282·9 \\
\hline
\text{Mean Area} = \quad 380·5
\end{array}
$$

$$
\begin{array}{r}
412·7 \\
(5 \text{ to } 7) \quad 720·5 \\
2223·1 \\
\hline
6)3356·3 \\
\hline
\text{Mean Area} = \quad 559·4
\end{array}
$$

$$
\begin{array}{r}
720·5 \\
(7 \text{ to } 9) \quad 844·8 \\
3124·8 \\
\hline
6)4690·1 \\
\hline
\text{Mean Area} = \quad 781·7
\end{array}
$$

General Mean Area computed by Hutton's Rule,

$$\frac{A + 4B + 2C}{6}$$

Tabulated for the numerator by successive additions—*equivalent to multiplication.*

1	. .		232·5
2	. .		1151·9
3	. .	{	349·2
			349·2
4	. .		1521·0
5	. .	{	412·7
			412·7
6	. .		2223·1
7	. .	{	720·5
			720·5
8	. .		3124·8
9	. .		844·8

1 to 9

$$
\begin{array}{r}
6)\,12062·9 \\
\hline
1 \text{ to } 9 = \quad 2010·5 \quad Gen.\ Mean \\
Area.
\end{array}
$$

Separate Mean Areas.

$$
1 \text{ to } 9 \quad . . . \left\{
\begin{array}{r}
288·9 \\
380·5 \\
559·4 \\
781·7
\end{array}
\right.
$$

Same as above = 2010·5

4

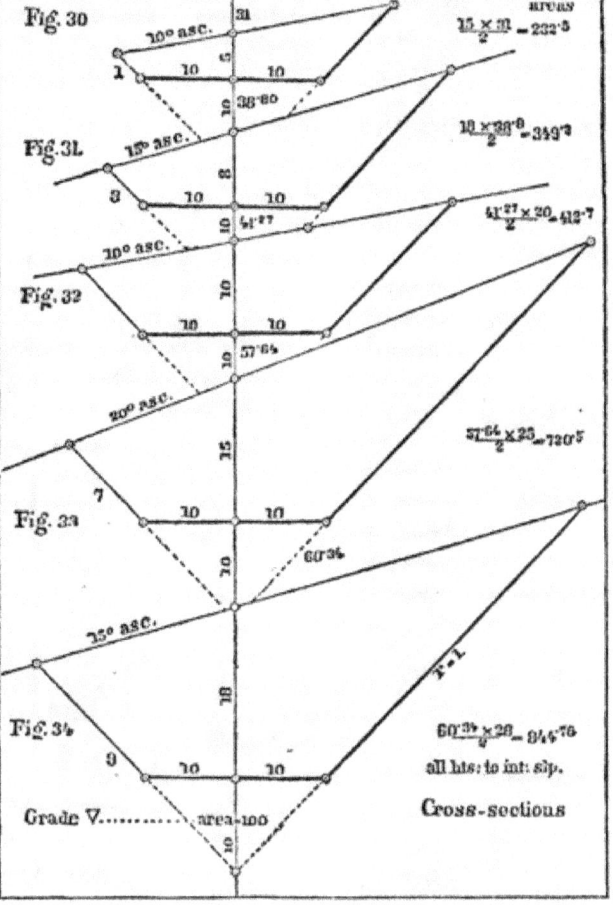

Mean areas computed separately for each regular station, by Simpson's Rule.

$$
\begin{array}{rl}
& 844\cdot8 \\
(9 \text{ to } 11) & 1085\cdot0 \\
& 3844\cdot0 \\
\hline
& 6)\overline{5773\cdot8} \\
\text{Mean Area} = & 962\cdot3
\end{array}
$$

$$
\begin{array}{rl}
& 1085\cdot0 \\
(11 \text{ to } 13) & 901\cdot5 \\
& 3964\cdot0 \\
\hline
& 6)\overline{5950\cdot5} \\
\text{Mean Area} = & 991\cdot8
\end{array}
$$

$$
\begin{array}{rl}
& 901\cdot5 \\
(13 \text{ to } 15) & 516\cdot0 \\
& 2781\cdot5 \\
\hline
& 6)\overline{4199\cdot0} \\
\text{Mean Area} = & 699\cdot8
\end{array}
$$

$$
\begin{array}{rl}
& 516\cdot0 \\
(15 \text{ to } 17) & 259\cdot5 \\
& 1506\cdot2 \\
\hline
& 6)\overline{2281\cdot7} \\
\text{Mean Area} = & 380\cdot3
\end{array}
$$

General Mean Area computed by Hutton's Rule.

$$\frac{A + 4B + 2C}{6}.$$

Tabulated for the numerator by successive additions—*equivalent to multiplication.*

1 to 17

$$
\begin{array}{rl}
\text{Br}'\text{t over } 1 \text{ to } 9 = & 12062\cdot9 \\
9 \quad . \quad . & 844\cdot8 \\
& 3844\cdot0 \\
11 \quad . \quad . \left\{ \begin{array}{r} 1085\cdot0 \\ 1085\cdot0 \end{array}\right. \\
& 3964\cdot0 \\
13 \quad . \quad . \left\{ \begin{array}{r} 901\cdot5 \\ 901\cdot5 \end{array}\right. \\
& 2781\cdot5 \\
15 \quad . \quad . \left\{ \begin{array}{r} 516\cdot0 \\ 516\cdot0 \end{array}\right. \\
& 1506\cdot2 \\
17 \quad . \quad . & 259\cdot5 \\
\hline
6)\; & 30267\cdot9 \\
\text{Gen. Mean Area} = & 5044\cdot7
\end{array}
$$

Separate Mean Areas.

1 to 17

$$
\begin{array}{rl}
\text{Brought over} = & 2010\cdot5 \\
& 962\cdot3 \\
& 991\cdot8 \\
& 699\cdot8 \\
& 380\cdot3 \\
\hline
\text{Total} \quad . \quad . = & 5044\cdot7
\end{array}
$$

(*Same as above.*)

Then, Mean Area.

$$\frac{5044\cdot7 \times 100}{27} = \underset{\text{C. yards.}}{18684\cdot1}$$

Deduct Grade Prism
for 8 stations =
$370\cdot4 \times 8$ . . . = $\;\;2963\cdot2$

Solidity. . . . . . = $\overline{15721\cdot}$
in cubic yards from
1 *to* 17.

So that the final solidity of this cut (as shown) from grade to ground, *vertically*, and from 1 to 17 (8 stations), *horizontally* = 15721 cubic yards (excluding for the present the grade passages).—A com-

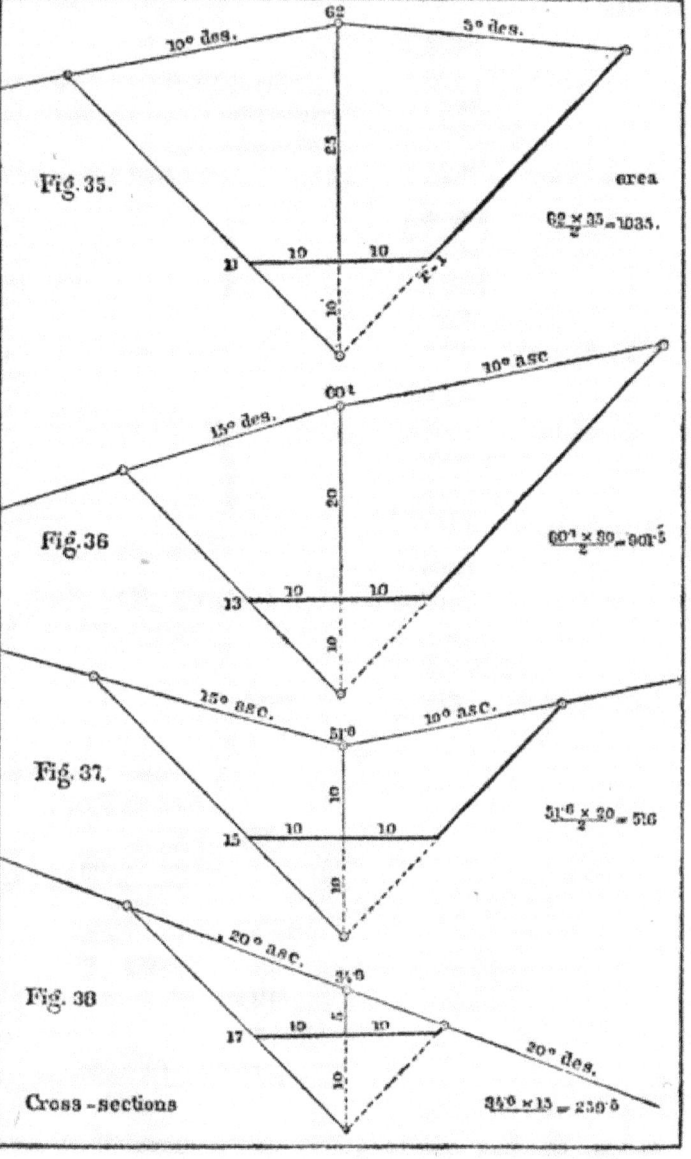

Fig. 35.

area

$\frac{62 \times 35}{2} = 1035.$

Fig. 36

$\frac{60^3 \times 30}{2} = 901\cdot5$

Fig. 37.

$\frac{51\cdot6 \times 20}{2} = 516$

Fig. 38

Cross-sections

$\frac{34\cdot6 \times 15}{2} = 259\cdot5$

parison of the calculated work, by Separate Mean Areas, and by General Mean Area,—*while resulting alike*, evinces the superiority of the latter, in point of *brevity*.

In the tabulation for General Mean Area, it will be observed that the extreme end areas are written but *once* (equivalent to addition) —the odd numbered areas *twice* (equivalent to $\times$ by 2), while the even numbered areas are written, in effect, 4 times,—as *squares of sums* of adjacent representative hights, because in that shape *they each equal* 4 *times the area of the prismoidal mid-section*.

(e.) . . . . . . We must now consider the passages from excavation to embankment at both extremities of the cut, near the regular stations, 1 and 17, where it was assumed *to be truncated*, in order to simplify its computation.

Figs. 39 to 42 show these passages so clearly, in the assumed case, as to need little explanation.

On plain ground the line of passage $a c$ will often be so nearly normal to the centre that, having set the grade peg in the centre line at e (the entrance of the cut), we may place those for the edges of the road-bed (as a and c), at right angles in many cases, where the ground differs in level only a few tenths of a foot; the error being merely a change of some yards from excavation to embankment, which is quite immaterial, since their values differ little per cubic yard.

But where the ground is much inclined, in either direction, the grade pegs $a e c$ must be set on an oblique line, broken at e, if necessary.

Precise rules can scarcely be furnished for such cases, but the quantities being usually small, and the distances short, any of the ordinary methods may be safely employed.

In the case before us, we have made the computation from 17 to a, and from 1 to a, by the Arithmetical Mean, and for the parts from a to c as pyramids.

In this manner we have found the volume of excavation, at the passage at *Fig.* 39, to be . . . . = 321 cubic yards.
And at *Fig.* 41 . . . . = 622 " "
Total, in the whole length of the passages ——
(230 feet) . . . . = 943 cubic yards.

So that, finally, we have for *the solidity* of the entire railroad cut, under consideration, the following result:

$$\begin{cases} \text{From 1 to 17 (as before computed)} = 15721 \text{ cubic yards.} \\ \text{In the passages from excavation to} \\ \quad \text{embankment, at both ends (230} \\ \quad \text{feet long in all)} \ . \ . \ . \ . \ . = \underline{\quad 943 \quad} \text{`` \quad ``} \\ \textit{Whole solidity} \text{ of the cut from grade} \\ \quad \text{to grade, on both sides} \ . \ . \ . = 16664 \text{ cubic yards.} \end{cases}$$

We will now illustrate *the passages* from excavation to embankment, at both ends of *the cut* (shown in profile at *Fig.* 29.)

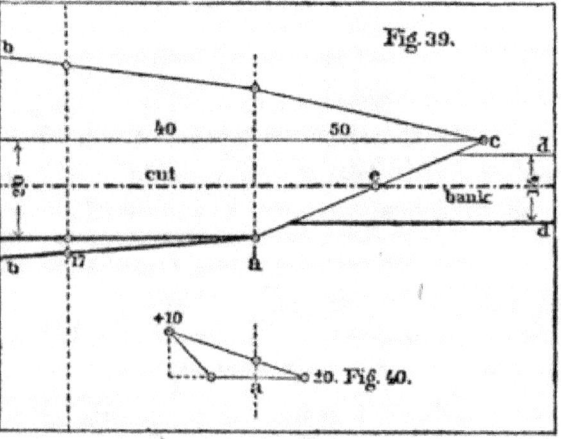

In *Figs.* 39 to 42 all letters refer to similar parts.

$$\begin{cases} \text{1 and 17} = \text{Places of cross-sections, at the selected regular stations,} \\ \qquad \text{where the cut } \textit{was truncated}, \text{ to obtain full work.} \\ a\,a = \text{Cross-section, where one edge of road-bed runs to grade.} \\ c = \text{Grade point at the other edge, or opposite side.} \\ a\,c = \text{Line of junction of cut and bank, at grade level.} \\ b\,b = \text{Slopes of cut.} \\ d\,d = \text{Slopes of bank.} \\ e = \text{Grade point at centre.} \end{cases}$$

Total length of cut between the extreme grade points forming the vertices of the small pyramids at c and $c = 1030$ feet.

Other modes may be used for treating the question of passages between excavation and embankment, but the above is as simple as any, and may be easily modified for particular cases.

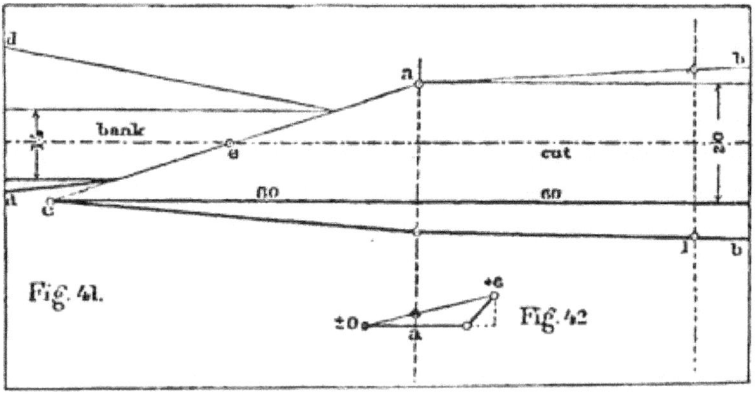

Fig. 41.

Fig. 42

11. *With Railroad Cross-sections in Diedral Angles—to find the mid-section of the Prismoidal Formula, by a brief calculation from the End Areas, without a Special Diagram.*

In all railroad cross-sections, instrumental data of adequate extent are first obtained in the field by well-known processes, and these data enable us in the office, subsequently, to draw them as diagrams, by a suitable scale, and to compute their superficies.

The length of each separate solid of earthwork, and its position upon the centre or guiding line, is also known.

With these given data, *the Prismoidal Formula* requires the deduction of a hypothetical mid-section, in some form, for use under the general rule, or its modifications.

As mentioned previously, this mid-section is usually derived from the Arithmetical Average of like parts in the end sections, and even in extremely irregular ground, to find this leading section of an Earthwork Prismoid, is not very difficult—when the diagrams of the end cross-sections are correctly drawn—(as in heavy work they always should be), or even from the field notes of the engineer, since the position of every leading point of ground, transversely, is always fixed and recorded by level hights, and distances out from centre, and their average position is always reproduced, *proportionally,* in the mid-section.

Nevertheless, some judgment is required in deducing the mid-sections from the end ones, by Arithmetical Means, since the points to average upon are often in doubt,—the process, too, including finding its area, is like most others connected with earthwork computations, very often tedious, so that some shrewd mathematicians, while conceding the accuracy of this method, when properly carried out, have, nevertheless, deemed it unsatisfactory in some respects.*

It is well, therefore, to have the means of operating with given end areas, to find the mid-section, without the necessity of arithmetically deducing, or even of sketching it.

We, therefore, now submit some rules and examples by which the area of the mid-section may be computed from the ends, without deriving it in the usual way, or drawing for it a special diagram.

These rules are intended only for Earthwork Prismoids, within diedral angles; and though their range is clearly more extensive, the variety of prismoidal solids is so great *that it is probably best to limit our rules and examples to the object before us.*

The broken ground line of very irregular cross-sections should always be reduced to a uniform slope, by a single equalizing line (or at most by two), containing *exactly* the same superficies, by the method of *Art.* 8,—and the hights and widths ascertained for each section (by the equalizing line), and verified by multiplication to re-produce the area equalized,—see 8 (a),—these hights and widths enable us at once to compute the volume of the prismoid by Simpson's Rule (their product giving end areas)—(*Art.* 2 (a))—and the sums of these hights and widths, when multiplied together, producing always 8 times the mid-section (without directly deducing it).

Having given then the end areas, or the hights and widths which produce them, we readily find *the Prismoidal Mid-section* by the following:

Rules.

$$(1.)\ \frac{\text{Arithmetical Mean} + \text{Geometrical Mean}}{2}\ .\ = \textit{Mid-sec.}$$

$$(2.)\ \frac{(\text{Sum of square roots of end areas})^2}{4}\ .\ .\ .\ = \textit{Mid-sec.}$$

$$(3.)\ \frac{\dagger\,\text{Sum end hights} \times \text{sum end widths}}{8}\ .\ .\ = \textit{Mid-sec.}$$

(4.) *By the method of Initial Prismoids—Art.* 3 (a).

---

* Warner's Earthwork (1861).—Davies' New Surveying (1870).

† These hights and widths (used in 3) are those connected with the equalizing line of the *equivalent* triangular section—the product of which, at each cross-section, re-produces *exactly* the double area of the whole surface, from the side-slopes to the broken ground line; and the product of their *sums* always equals *eight times the mid-section.*

Other rules might be given, but these *four* appear to be the simplest and best for use in earthwork, under the view we have herein taken.

Having then found the mid-section, and having the end areas and length previously given, we can easily compute the volume of any earthwork solid, by *the Prismoidal Formula*, or its numerous modifications.

By Geometry, we have for the *mid-sections* of . .
$$\begin{cases} 1. \text{ A Prism} \ \ldots \ \ldots \ \ldots \ = \text{ Base.} \\ 2. \begin{cases} \text{A Wedge, with back} \\ \text{ and edge equal and} \\ \text{ parallel} \ \ldots \ \ldots \end{cases} = \tfrac{1}{2} \text{ Base.} \\ 3. \text{ A Pyramid} \ \ldots \ \ldots \ = \tfrac{1}{3} \text{ Base.} \end{cases}$$

Fig. 43 shows the end cross-sections of one station of a railroad cut, upon irregular ground, *both upon one diagram*, road-bed 20, side-slopes 1 to 1.   Length of station, 100 feet.

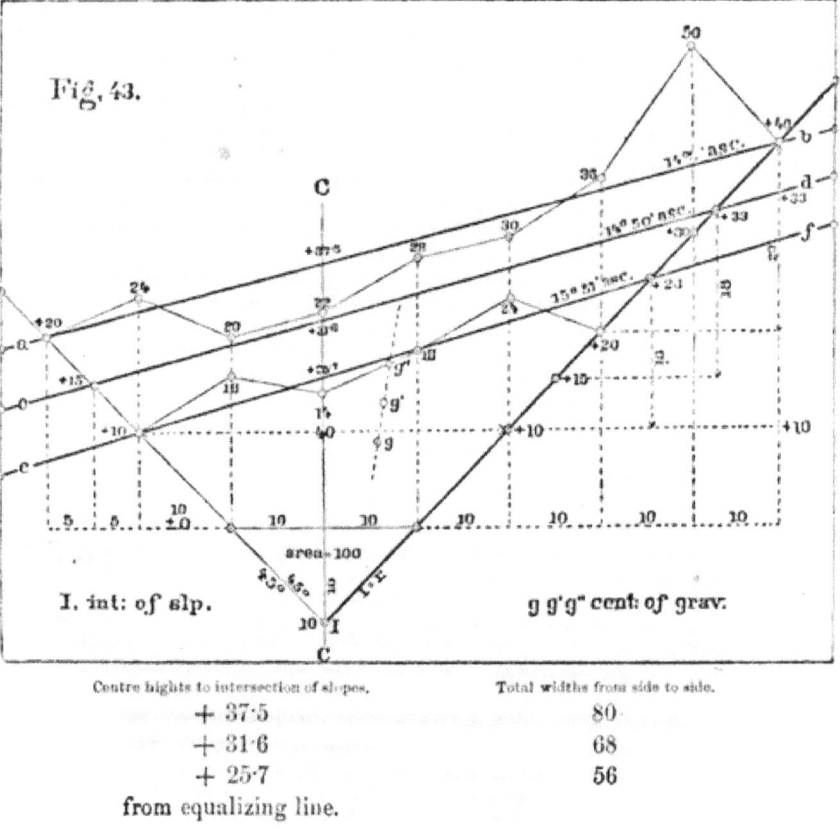

Fig. 43.

I. int: of slp.   g g'g'' cent: of grav.

Centre hights to intersection of slopes.	Total widths from side to side.
+ 37·5	80
+ 31·6	68
+ 25·7	56

from equalizing line.

Note:

Both in *Figs.* 43 and 44 the same letters refer to like parts.

CC = Centre line of railroad, or guiding line of earthwork.

ab = Equalizing line of broken ground surface of larger end . . = 14° 2′ slope,

ef =   "   "   "   "   "   of smaller end . . = 15° 57′ "

ad =   "   "   "   "   "   of mid-section . . = 14° 50′ "

Fig. 44, like the preceding, shows both end sections of a railroad cut, *upon one diagram.* Road-bed = 20, side-slopes 1½ to 1. Length = 100.

Centre hights to intersection of slopes.	Total widths from side to side.
+ 22·02	66·
+ 26·07	78·7
+ 29·81	90·7

from equalizing line.

In this figure (44) the line ef has a minus slope, which is always the case when the area assumed up to the equalizing point is greater than *that to be equalized.*

In both of the above figures, I is the intersection of the side-slopes, or edge of the diedral angle, containing the earthwork prismoids.

The constant area of the grade triangle, with side-slopes of 1 to 1 (*Fig.* 43) = 100. While, with side-slopes of 1½ to 1 (*Fig.* 44) = 66⅔. The road-bed, or graded width, in both cases being 20 feet. The altitude of this triangle for 1 to 1 = 10, and for 1½ to 1 = 6⅔.

The rules (numbered) above, for the figures *shown*, give the following results:

{ *Fig.* 43 *gives Mid-sections* (1) = 1074·5; (2) = 1074·5 ; (3) = 1074·4 ; (4) = 1074·5
{ *Fig.* 44 *gives Mid-sections* (1) = 1015· ; (2) = 1014·74; (3) = 1015·22; (4) = 1015·

The small variations arise from the decimals not being sufficiently extended.

12. *To find the Prismoidal Mean Area from the Arithmetical or Geometrical Means, or the Mid-section, by Corrective Fractions of the Square of the Difference of End Hights.*

In all cases we suppose the *end areas* of the Prismoid *to be given*, and that the Prismoid itself is contained *within a diedral angle*, the plane angle measuring it being supplemental to double the angle of side-slope, as in the *Figs.* 43 and 44.

The simplest, and probably by far the most generally employed method of finding a mean area between two others,—is by the Arithmetical Mean—which is itself *half the sum of any two magnitudes.*

Adopting the Arithmetical Mean as being the simplest known base, and forming all sections of earthwork by prolonging the planes of the side-slopes to their intersection (or supposing them to be), so as to bring the computed prismoids within diedral angles of given divergency.

We have, from the relations between the sums or differences of the squares, or rectangles of lines producing areas, some rules, which may often be useful in the calculation of earthwork, for correcting mean areas to be used in finding *the solidity.*

This correction being always equivalent to some fraction of the square of the difference of the end hights.

While these end hights are always to be deemed and taken as the square roots of the end areas, and are, in fact (as before mentioned), a side of an equivalent square, or half base of an equivalent triangle, having side-slopes of 1 to 1 (or a diedral angle of 90°),—for (*we repeat*), no matter what may be the ratio of actual side-slope, nor how irregular the ground surface, *the square root of the area* is invariably the true representative hight which rectifies the section, *and which, when squared, reproduces the area.*

See *Art.* **10** (**a**) (**b**) etc., where much use is made of these square roots, or representative hights.

Having, then, the end areas given, and their square roots or hights ascertained,

$$D = \text{Difference of hights.}$$
$$D^2 = \text{The square of the difference of hights.}$$

Rules:

(1) *Arithmetical Mean* $= \dfrac{\text{Sum end areas}}{2}$.

Then the Prismoidal Mean Area.

(2) . . $=$ Arithmetical Mean $- \frac{1}{6}\,D^2$.

(3) . . $=$ Mid-section . . . $+ \frac{1}{12}\,D^2$.

(4) . . $=$ Geometrical Mean $+ \frac{1}{3}\,D^2$.

Prismoidal Mid-section.

(5) . . $=$ Arithmetical Mean $- \frac{1}{4}\,D^2$.

Geometrical Mean.

(6) . . $=$ Arithmetical Mean $- \frac{1}{2}\,D^2$.

For *Fig.* 43 these rules give,

(1) $= 1110\cdot$ $=$ Arith. Mean.
(2) $= 1086\cdot4$
(3) $= 1086\cdot3$ $\Big\}$ $=$ Pris. Mean.
(4) $= 1086\cdot4$
(5) $= 1074\cdot6$ $=$ Pris. Mid-sec.
(6) $= 1039\cdot2$ $=$ Geom. Mean.

For *Fig.* 44 these rules give,

(1) $= 1039\cdot$ $=$ Arith. Mean.
(2) $= 1022\cdot9$
(3) $= 1023\cdot$ $\Big\}$ $=$ Pris. Mean.
(4) $= 1023\cdot2$
(5) $= 1014\cdot8$ $=$ Pris. Mid-sec.
(6) $= 991\cdot$ $=$ Geom. Mean.

In these numerical illustrations (as in others) slight variations arise from insufficient decimals.

Baker* gives yet another rule for the Prismoidal Mean Areas, as follows:

$$\frac{\text{Sum end areas} + \text{Rectangle hights}}{3} = \text{Prismoidal Mean.}$$

And we may repeat, as another modification of the *Prismoidal Formula*, arising from this discussion, the following (same as **XI.**, before given):

XII. . . . . *Solidity*

$$= \frac{(\text{Sum of squares of hights}) + (\text{Square of sum of hights})}{6} \times h.$$

---

* Baker's Railway Engineering and Earthwork (London, 1848). Other writers have given the same, and it is deducible from Hutton's Mens., Prob. 7, *as most of these Formulas are.*

This is equivalent to $\dfrac{2\,(\text{Sum sqs.}) + 2\,(\text{Rect. hights})}{6}$, or $\div\,2 =$

$\dfrac{(\text{Sum of sqs.}) + (\text{Rect. hights})}{3}$, which is Baker's rule above, or *Bid-der's*, as quoted by Dempsey (Practical Railway Engineering (4th edition) 1855).

We may illustrate this matter further by two simple figures.

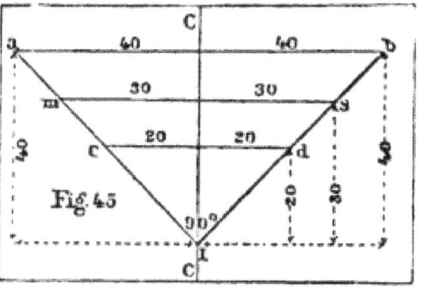

Fig. 45

Here *Fig.* 45 represents a 1 to 1 side-slope—diedral angle 90°; and *Fig.* 46 a side-slope of 1½ to 1—diedral angle 112° 38'.

In both these diagrams the same letters refer to like-parts.

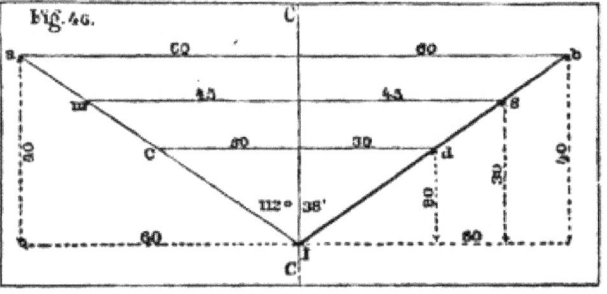

Fig. 46.

References.

CC = Centre line.
I   = Intersection of planes of side-slope.
a b = Ground line of one end section.
c d =   "   " , of the other.
m s =   "   " of the mid-section.
Hights and areas both extend to the intersection at I.

In Fig. 45, The end areas are 1600 and 400—the hights 40 and 20—and by the rules herein, Arithmetical Mean = 1000, Geometrical Mean = 800, Mid-section = 900, Prismoidal Mean Area = 933⅓, *by all the rules.*

In Fig. 46, The end areas are 2400 and 600—the hights = 48·99 and 24·99, being the square roots of the respective end areas—and by the rules herein, Arithmetical Mean = 1500, Geometrical Mean = 1200, Mid-section = 1350, Prismoidal Mean Area 1400, *by all the rules.*

The areas and hights, in both examples, are contained between the ground lines, and the intersection of the planes of side-slope, or edge of diedral angle, *including the Prismoid of Earthwork.*

13. *Applicability of the Prismoidal Formula to find the Solidity of Various Solids other than Prismoids.*

The Prismoidal Formula appears to be the *fundamental rule* for the mensuration of *all* right-lined solids, and the special rules given, in works on mensuration, for ascertaining the volume of solids in general use, seem like mere cases of the former; though their relation has never been demonstrated in plain terms by mathematicians—so as to con-nect them *directly*—further than *prisms, pyramids,* and *wedges,* which has already been done by the present writer in Jour. Frank. Inst., 1840.

Nevertheless, Hutton (1770) has indicated numerous applications, and various writers have since shown the applicability of *the Pris-moidal Formula* to ordinary solids, and also its coincidence with many special rules of the books, when proper algebraic substitutions are made; and it has been further shown to hold for certain warped solids, to which its application *was not expected.**

As an evidence of its remarkable flexibility, we may show, briefly, its application to *the three round bodies,* illustrated by a diagram.

(1) *The volume of a cone equals the product of its base ✕ ⅓ its hight.†* The prismoidal mid-section of a cone = ¼ the area of the base. The section at the top, or vertex = 0. Then, the sum of these areas used *prismoidally* = 2 base, which, ✕ ⅙ h = base ✕ ⅓ hight, which is the geometrical rule.

---

* Gillespie, Frank. Inst. Jour. (1857 and 1859).—Warner's Earthwork (1861).

† Chauvenet, ix. 3, 7, 14, Geom. (1871).—Borden's Useful Formulas (1851).—Henck's Field Book (1854), Art. 112.

(2) *The volume of a sphere equals 4 great circles × ⅓ its radius.** Now, the prismoidal sections at the poles are both = 0. While four times the mid-section = 4 great circles. Then, the *prismoidal* sum of areas = 4 great circles, which × ⅙ hight, or diameter, or ⅓ radius, is the geometrical rule.

(3) *The volume of a cylinder equals the product of its base by its hight.** Now, by the Prismoidal Formula, base + top + 4 times mid-section = 6 base (for all the sections are alike), and 6 base × ⅙ h = base × hight, which is the geometrical rule.

So that there can be no doubt of the applicability of the Prismoidal Formula *to the three round bodies;* and in a similar manner it is easy to show its coincidence with many special rules for solids, but a *direct* mathematical demonstration connecting all these together, and exhibiting their geometrical relations, has never come under the writer's notice; though *indirectly*, and perhaps quite as satisfactorily, this connection has been clearly established *for all the leading solids in practical use.*

Numerical calculation of the three round bodies, supposing each to have a diameter of 1, and an altitude of 1.

CONE.		SPHERE.		CYLINDER.	
Prismoidally.	Geom. Rule.	Prismoidally.	Geom. Rule.	Prismoidally.	Geom. Rule.
Top . . = ·0		Top . . = ·0	4 great circles	Top . . = ·7854	
Mid. × 4 = ·7854	Base = ·7854	Mid. × 4 = 3·1416	= 3·1416	Mid. × 4 = 3·1416	Base . = ·7854
Base. . = ·7854	1 × ⅓	Base. . = ·0	× ⅓ of ½	Base. . = ·7854	1
6)1·5708	*Solidity* = ·2618	6)3·1416	*Solidity* = ·5236	6)4·712	*Solidity* = ·7854
·2618		·5236		× 1	
1		1		*Solidity* = ·7854	
Solidity = ·2618		*Solidity* = ·5236			
Ratios of volume 1		.... 2		 3	

ab = The Base.
cd = " Top.
ms = " Mid-section.

Fig. 47.

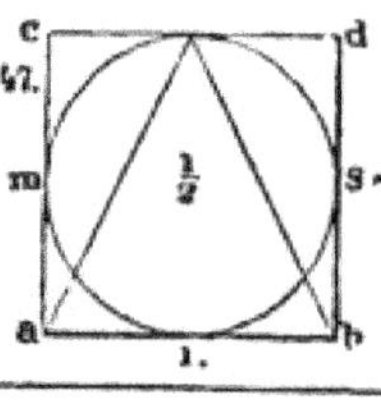

The common rules of mensuration are drawn from geometry—but geometry also teaches that *a cone*, *a sphere*, and *a cylinder*, dimensioned and situated as shown by their right sections, in *Fig.* 47, have

* Chauvenet, ix. 3, 7, 14, Geom. (1871).—Borden's Useful Formulas (1851).—Henck's Field Book (1854), art. 112.

their volumes in the ratio of the numbers 1, 2, and 3.—Now, the above calculations show the same result numerically, which, with the preceding observations, furnish an adequate demonstration.

In like manner we might show that *the Prismoidal Formula* applies to all the separate geometrical solids, which, when aggregated, form the irregular prismoid known as *an Earthwork Solid.*

Now, considering this species of solid as a prismoid, within the limits of Hutton's definition (1770), we find that all such admit of decomposition into Prisms, Prismoids,* Pyramids, or Wedges (*complete or truncated*), or some combination of them, having a common length, or hight, equal to the distance between the end areas or cross-sections, *and either separately or together computable by the Prismoidal Formula as a general rule for all.*

By a similar analogy (to the three round bodies), we find somewhat like relations to obtain between what we may call *the three square or angular bodies;* which geometry shows to exist alike amongst them all, the round bodies being referred *to the cylinder;* the square or angular ones *to the cube.*—But the wedge requires this special definition, that the edge be *double* the back.

1. *A Pyramid*, with a square base, on a side of 1, and
having also an altitude of 1, has a volume . . . . $= \frac{1}{3}$.

2. *A Wedge, doubled on the edge*, with a square back, on a
side of 1, the edge parallel $= 2$ (or double the back),
and an altitude of 1, has a volume . . . . $= \frac{2}{3}$.

3. *A Cube, or Hexaedron*, with its six square faces, each
formed upon a side of 1, has a volume . . . . . $= 1$.

So that, finally, we have, both *in the three round, and in the three square bodies* (as defined) *where unity* is the controlling dimension, *like ratios of volume.*

Thus, these six bodies,

{ Cone and	Sphere and	Cylinder ⎫	Solids of
{ Pyramid.	Wedge	and Cube. ⎪	Circular
	(doubled on the edge).	⎬	and
Have the same ratios of volume $= 1$.	2.	3. ⎭	Square Bases.

And of each and all of these alike, *the Prismoidal Formula* gives the *Solidity.*

---

* The Rectangular Prismoid being always divisible into two wedges.

14. *Transformation of Areas into Equivalent ones, Simpler in Form, and of Solids into Equivalents, more readily Computable by the Prismoidal Formula, or its Modifications.*

Hutton hath defined a Prismoid *as follows:*

"A Prismoid is a solid having for its two ends any dissimilar plane figures of the same number of sides, and all the sides of the solid plane figures also." (Quarto Mens., 1770.)

This is the oldest and best definition of the Prismoid which we are able to find on record.*

Under this definition, for which the General Rule (coinciding with Simpson's) was framed by Hutton, it is clear that we ought not to expect of the Prismoidal Formula the cubature of curvilinear solids, though, by a happy coincidence, it applies to many such, which are not prismoids at all, nor in the least resemble them, *geometrically.*

But though often true of this remarkable formula, where a correct mid-section can be first obtained, it by no means follows that its numerous modifications (all framed for right-lined solids) will, like their principal, *also hold, as it does in many singular cases exactly, and in most others approximately.*

It was early discovered that it would materially simplify the computation of irregular prismoids, to transform them into equivalent right-lined bodies, of which the nature was better known, and the forms more regular and simple.

As the calculations for level ground were obviously the most easy, Sir John Macneill, in his Tables of 1833, adopted for the end sections the principle of transformation into level hights, to contain equivalent level areas—and was, in fact, the originator of what has since been known *as the Method of Equivalent Level Hights*—by means of which, the end sections of irregular prismoids of earthwork are transformed into level trapezoids, which are then employed to compute an equivalent solid of the same length, and transversely level, at top or bottom, according as it may be excavation or embankment—each, however, representing the other, *when inverted.*

Sir John Macneill has been followed, more or less closely, by most of the authors of Earthwork tables, the bulk of which are applicable to level ground alone, or ground reduced to such;—though Watner's System of Earthwork Computation (1861) deals with ground however sloping, *or even warped, within certain limits.*

---

* See also Henck's Field Book (1854).—Davies Legendre (1853).—Haswell's Mens. (1863).—Bonnycastle's Mens. (1807).—Hawney's Mens. (1798). *All define the Prismoid as a right-lined solid.*

The method of using Equivalent Level Hights (when the cross-section of the ground is not level) has been concisely explained, by a recent writer, to consist *in finding,*[*]

1. "The area of a cross-section at each end of the mass."
2. "The hight of a section, *level at the top*, equivalent in area to each of these end sections."
3. "From the average of these two hights, the middle area of the mass."

"And, *lastly*, in applying the Prismoidal Formula to find the contents."

It is obviously necessary then to understand what is meant by *equivalency*—and this we find from Geometry.[†]

1. "*Equivalent (plane) figures* are those which have the same surface—measured by the area."
2. "*Equivalent solids* are those which have the same bulk or magnitude."

"*Theorem:* If two solids have equal bases and hights, and if their sections made by any plane parallel to *the common plane* of their bases are equal, they are equivalent."

Now, the transformation of triangular prismoids of earthwork, by means of Equivalent Level Hights, meets every point of Professor Peirce's definitions of *equivalency*, and hence the solid they produce may be regarded as *equivalent* to the original defined by Hutton :—in the above theorem, equality of sections evidently means *equality in area*, and not geometrical equality, which is somewhat different.

Some writers have doubted the accuracy of the transformation or *equivalency* produced by Equivalent Level Hights,[‡] but it is because the solids, which they found in error, were either not prismoids at all, or else the data used were *inadequate* to the solution of the problem.

An error in this direction is not surprising ; for when we know that *the Prismoidal Formula* applies correctly to a solid, we are apt to infer that its modifications also do,—*and here the error lies.*

For instance, we know this formula *does apply* correctly to a sphere, but if we test *that solid*, by the method of Equivalent Level Hights, we should find that the end sections being 0, have a hight of 0, and that the mid-section being constructed on a mean of like parts in the

---

* Henck's Field Book (1854).   † Peirce's Plane and Solid Geom. (1837).

‡ Gillespie, Frank. Inst. Jour. (1859).

ends must also equal 0, and hence we might in this way legitimately come to the conclusion that the globe itself had a solidity of 0 ! This shows that Equivalent Level Hights are *limited* in range.

The error obviously is—that all, or most of the transformations and modifications of the Prismoidal Formula, are intended for right-lined solids, "*varying uniformly*" from end to end, like a stick of timber dressed off *tapering*, and to all such rectilinear solids they do apply correctly ; but not to those which bulge out, or curve in, *by laws unknown to Hutton's definition of the Prismoid.*

It would be easy to illustrate this by examples, and to show that, confined within proper limits, the usual modifications of the Prismoidal Formula are correct enough for practical use ; *but they have not the wide range of their principal;* nor must they be expected to apply either to the three round bodies, or to warped solids, *but only to right-lined ones, varying uniformly, or nearly so, from end to end.*

One important point, however, must not be overlooked in applying the Prismoidal Formula (or its modifications) to cases of earthwork : that is, *the ground must be properly cross-sectioned ;* or, have its sections judiciously located, while the hights and distances of its controlling points are correctly measured and recorded, prior to undertaking the calculations of *solidity.*

It is in this point that Borden's *ridge and hollow problem fails.** Had one or more intermediate cross-sections been adopted there, no difficulty would have existed in its calculation, either by Borden himself, or by subsequent students.

To illustrate this subject, we will give an example, drawn from Simpson's original Prismoid of 1750, on which he founded *the Prismoidal Formula,* or used to explain it. *Art.* 2, *Fig.* 2. (And see *Figs.* 48, 49, 50, 51.)

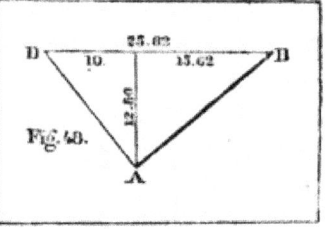

Fig. 48.

Here we will take the Prismoid as being cut *in two*, by the diagonal plane, through DB, so as to divide it into triangular prismoids, and then calculate one of these halves in three ways.

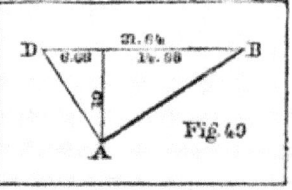

Fig. 49

* Borden's Useful Formulas, etc. (1851).—Henck's Field Book (1854).

1. By Simpson's Rule, as the half of a rectangular prismoid, *dimensioned* as in *Fig.* 2.

2. By Hights and Widths, as a triangular earthwork solid, with unequal side-slopes. (See *Figs.* 48, 49.)

3. By Equivalent Level Hights purely as an *equivalent* triangular prismoid, or earthwork solid, within a diedral angle of 90°, and having equal side-slopes of 1 to 1.

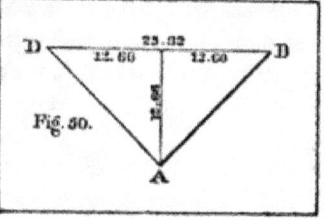

Fig. 50.

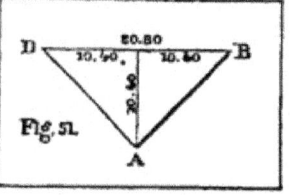

Fig. 51.

In all these figures the angle A = 90°.

B and B, *Figs.* 48 and 49 = 38° 40', and 33° 41'.

Areas, $\begin{cases} 48 \text{ and } 50 = 320. \\ 49 \text{ and } 51 = 216. \end{cases}$

The common hight of the prismoids being $h = 24$. All the calculations being carried out in detail; all having the same end areas, 320 and 216; and all *dimensioned* as marked upon the figures.

We find, then, by all these calculations, the *Solidity* to be the same = 3200, varying but a few small decimals, and agreeing with the results already ascertained in *Art.* 2.

This exhibits the *equivalency* we have been discussing (the figures being quite unlike), and might readily be extended to more complicated examples, *with a like result.*

15. *Equivalence of some important Formulas, for computing the Solidity of Triangular Prismoids of Earthwork, contained within Diedral Angles, formed by Prolonging the Side-slope Planes to an Edge.*

Equivalent Formulas are those which reach the same results by unlike steps—and in mathematical processes it is often found that a general formula will hold in many cases, usually governed by concise special rules, *and yet produce identical results.*

This is *equivalency*, and relates in mensuration especially to *the Prismoidal Formula*, which appears to have a sort of concurrent jurisdiction over the domain of solid geometry, along with the special rules for the volume of each separate solid, producing exactly the same results, though by different steps.

Such is particularly the case in earthwork solids, contained (as they mostly are) in diedral angles formed by uniform planes, called side-slopes, and having a general *triangular* section—two sides being the inclined lateral planes, known as side-slopes (continued to intersect for computation), and these slopes being usually alike in inclination, while the contained angle is equal;—the third side, *or ground line, alone being variable, and often irregular.*

By geometry, triangles having an angle common or equal, and the containing sides proportional, *are similar;* and the areas of similar triangles are always proportional to the squares of any similar or homologous lines, or to the rectangles of such as have like positions and relations to each other :—as the squares of perpendiculars from the equal angles, or their bisectors, the rectangles of containing sides, the product of hights and widths, etc.

Now, these triangular sections of an earthwork solid, extending (for computation) from the ground surface to the intersection of the side-slopes prolonged to an edge, *are sections of triangular pyramids, as well as of prismoids;* and to such solids the rules for Pyramids, and their frusta, as well as the Prismoidal Formula, and its modifications, apply *concurrently,* and either may be used at will, with correct results.

These considerations regarding the equivalency of *Pyramidal* and *Prismoidal* Formulas in such cases are important, and require to be well considered by computers of earthwork.

Hutton's definition of the Prismoid is based on three conditions:

1. The two ends must be *dissimilar* parallel plane figures.
2. They must have *an equal number of sides.*
3. The faces, or sides of the solid, *must be plane figures also.*

Usually, says Hutton, *the faces are plane trapezoids.*

Considering, now, a regular prismoid as being composed of known elementary solids.

Macneill regards it as formed of a prism, with a wedge superposed. *Art.* 4 (and this is also the case with a frustum of a pyramid, turned upon its edge).

Hutton, of two wedges, formed by a single cutting plane passed in a diagonal direction, *Art.* 3.

The writer, as a triangular prism *trebly truncated, Art.* 1.

Simpson (the father of the prismoid) gives no special definition, but figures in his work of 1750 *a rectangular prismoid* (the same or similar to that adopted and figured by Hutton, 1770); and by a single diagonal plane, convertible into two triangular prismoids. (See *Fig.* 2.)

Now, as a triangle is the simplest of all polygons, so a prismoid within a diedral angle (triangular in section) may be considered as the simplest of all prismoids, though the rectangular prismoid is nearly so.

The simplest case of the ordinary trapezoidal prismoid of earthwork is in, or upon, *ground level transversely.*

In that case, the cross-sections *are level trapezoids,* and the solid is obviously composed of a prism and superposed wedge, as in Macneill's solid, *Art.* **4.**

Its volume may be computed by Simpson's, or by Hutton's general rules, because this solid then is strictly a prismoid within the scope of Hutton's definition, and as a whole computable *only* by prismoidal rules.

But suppose the assumed road-bed was taken less and less, until we reached the edge of the diedral angle, and it became *zero.*

Then, the cross-section from a trapezoid becomes a triangle, and the prismoid changes at once into a frustum of a pyramid—a solid known since the days of Euclid.

This solid becomes then computable by Euclid's geometry, as the frustum of a pyramid—or by Equivalent Level Hights—by roots and squares—by geometrical average—all of which are equivalent, as are the similar rules of Bidder, Baker, Bashforth, and others; or, by wedge and prism, by hights and widths (Simpson), by Hutton's particular rule, by the method of initial prismoids, or, finally, by *the Prismoidal Formula itself,* which always holds *alike* for prismoids, pyramids, or pyramidal frusta.

Hutton (4to Mens., 1770, p. 155) shows that in similar sections of a pyramidal frustum (say triangular) the squares of similar lines, as the bisector of an equal angle (which the centre line of a railroad generally is), are as the areas of the cross-sections, or, conversely, the areas are as the squares of similar lines (Chauvenet's Geom. iv. 7).

Then, from Hutton's prob. 7, cor. 2, we have a formula (for pyramidal frusta) in which, substituting Bidder's and Baker's notation, we have, by a slight reduction, *the identical rules* given by those authors for the computation of earthwork.[*]

---

[*] Bidder, quoted in Dempsey's Prac. Rail. Eng., London, 1855.—Baker, in his Railway Eng. and Earthwork, London, 1848.

We will now give a diagram to illustrate *the equivalency* of prismoidal and pyramidal formulas.

Fig. 52.

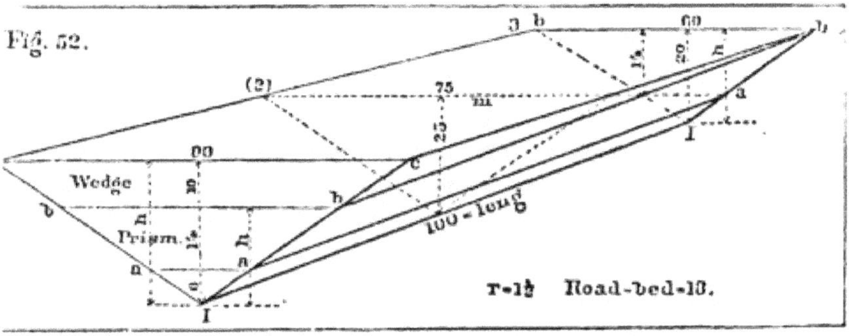

Fig. 52 represents the full station of earthwork, already shown in *Figs.* 22 and 24, having a road-bed of 18 feet, and side-slopes of 1½ to 1, with other dimensions as marked upon the figures.

Suppose, *in all cases* (as in *Fig.* 52), the trapezoidal sections of the ends above the road-bed to be carried down by prolonging the side-slopes to their intersection at I I, the edge of the diedral angle.

Let $\begin{cases} e\,e &= \text{Top of larger end, and } h = \text{ its hight} = 30 \text{ feet.} \\ b\,b &= \text{Top of smaller end, and } h' = \text{ its hight} = 20 \text{ feet.} \\ I &= \text{The intersection of side-slopes, of } 1\frac{1}{2} \text{ to } 1. \end{cases}$

Then, suppose a horizontal plane to be passed parallel to I I, through $b\,b\,b\,b$, then $c\,c\,b\,b\,b\,b$, the part cut off, *is a wedge,* its edge being $b\,b$, the top of the forward cross-section; while $h - h' = $ the hight of the back $c\,c\,b\,b$,—and as *a wedge* it may easily be calculated.

Now, suppose the plane $b\,b\,b\,b$ moves *downward*, parallel always to its first position at the distance h' from I, then the solid immediately becomes a prismoid—being then a prism with a wedge superposed, as in *Art.* 4 (or analogous to it).

Continue this parallel movement of the plane *downward* until we reach the position $a\,a\,a$, assumed for the road-bed, and then we have the precise case of *Art.* 4—Sir John Macneill's figure of 1833. To this of course *the Prismoidal Formula* applies, but the Pyramidal Formulas *do not.*

Continue on again, with the movement of our supposed horizontal plane *downwards*, until it comes to I, I, (the junction of the side-slopes), then the solid becomes the frustum of a pyramid, triangular in section, *and the wedge is absorbed ;* nevertheless, a frustum of a pyramid is also in this respect like unto a prismoid, and may, if we choose, be regarded as a prism with a wedge superposed, and forming the top of the solid.

Taking the horizontal plane, supposed to move parallel downwards, at three particular points of its progress,—at *b*, *a*, and I,—the calculations for volume would be,

1. For the *wedge* alone $= c\,c\,b\,b\,b\,b$
2. "   wedge and prism, *or prismoid* $= c\,c\,a\,a\,a\,b\,b.$
3. "   *frustum of a pyramid* alone, both wedge and prism being merged in it—and in such case this is the simplest and best form of calculation, *for volume.*

We may here remark that so long as the end cross-sections contain a road-bed of definite width, the solid is a real prismoid, and must be computed as such *by prismoidal rules alone;* but the moment the angle at I becomes common to both, then the solid becomes a regular frustum of a pyramid, and all the pyramidal rules apply, as well as the prismoidal ones, *to which they are strictly equivalent,* whenever I, the diedral edge, is common to both.

Now, suppose the case *reversed*, and that the horizontal plane was originally passed through I, I, (edge of diedral angle), and moves gradually *upwards*, parallel.

At every step of its progress, the solid, cut off above I, is always a prism, until *its limit* has been reached, at *b b b b*, the top of the smaller end—here the moving horizontal plane ceases to be longer useful in illustration; and becoming fixed at one end, on the top of the *far end* section as an axis, opens wider and wider at the *near end*, until it attains the line *c c* (the top of the main solid), and completes the wedge we have referred to, *and the pyramidal frustum with it.*

In this position the whole solid is *undeniably a prismoid* (if we allow to it an infinitesimal road-bed). So, also, it is *a frustum of a triangular pyramid*, both being *strictly equivalent*, and both computable *by the regular rules for either.**

We will now illustrate this equivalence of the *Prismoidal and Pyramidal Formulas*, in their application to earthwork solids, within diedral angles, by a few examples.

Taking the dimensions of *Figs.* 22 and 24, with 1½ to 1 side-slopes, and road-bed of 18, for the numbers to be employed—the diedral angle being common to both.

---

* As might be inferred from Hutton's remarkable chapter on the Cubature of Curves (4to Mens., 1770).

1. *Prismoidally.*—By the direct and cross multiplication of Hights and Widths.   Formula at the end of *Art.* **9.** . . . . . . **VIII.**

$$\text{Hights} \begin{cases} h = 30 \\ h' = 20 \end{cases} \times \begin{cases} w = 90 \\ w' = 60 \end{cases} \text{Widths.}$$

30	20	30	90	2700
90	60	60	20	1200
2700	1200	2)1800 + 1800		1800
		1800		6)5700

$$950 \times 100 = 95000 =$$
Solidity, as before computed.

2. *Pyramidally.*—By the rules of Baker's Earthwork.

30	20	30	900
30	20	20	400
900	400	600	600
			1900

$r = 1\frac{1}{2}$ 50
$l = 100$

3)150
50

$95000 = $ *Solidity,* as before computed

3. *Prismoidally.*—By Simpson's rule, modified for triangular solids.

Hights.		Widths.		
	30	×	90	= 2700
	20	×	60	= 1200
Sums,	50	×	150	= 7500

12)11400

$$950 \times 100 = 95000 = \text{\textit{Solidity,}} \text{ as}$$
before computed.

4. *Pyramidally.*—By Roots and Squares, *Art.* **10** (**c**).

End Areas . . = 1350   600
Roots. . . . = 36·74   24·50
Sum . . . . = 61·24
Square of Sum = 3750
End Areas . . = { 1350
                 { 600

6)5700

$$950 \times 100 = 95000 = \text{\textit{Solidity,}} \text{ as}$$
before computed.

5. *Finally, by Warner's Earthwork, Art.* 112.

$$\text{Difference} = 10 \begin{cases} \overset{\text{Hts.}}{30} \times \overset{\text{Wds.}}{90} \\ 20 \times 60 \end{cases} \text{Difference} = 30.$$

$$\text{Sums} \quad . \quad . \quad 50 \times 150 \quad . \quad . \quad . \quad . \quad . = 7500$$

$$\div 8 = \overline{937\cdot5} = \text{1st term.}$$

$$\frac{10 \times 30}{8 \times 3} = \frac{12\cdot5}{950} = \text{2d term.}$$

$$\times 100 = \overline{95000} = \textit{Solidity.}$$

So, we may safely assume that the *Pyramidal Formulas* of Bidder, Baker, and others, the Geometrical Average, Equivalent Level Hights, Euclid's rule for the frustum of a pyramid, etc., *are all* strictly equivalent *to the Prismoidal Formula*, and its modifications, when applied to earthwork solids, *within diedral angles,*—on ground transversely level.

16. *Summary of Rules and Formulas from the Preliminary Problems.*

It will be found convenient to use, substantially, the same notation for the Prismoidal Formula, and its numerous modifications, wherever practicable.

$$\textit{Thus let} \begin{cases} b = \textit{Base,} \text{ or area of end assumed for such.} \\ t = \textit{Top,} \text{ or area at the other end.} \\ m = \textit{Hypothetical Mid-section,} \text{ used in computation.} \\ h = \textit{Length or hight} \text{ of the Prismoid.} \\ S = \textit{Solidity or volume.} \end{cases}$$

Then, the Prismoidal Formula can always be in substance expressed by $\dfrac{b + t + 4m}{6} \times h = S$, when a mean area is desired, or by $(b + 4m + t) \times \frac{1}{6} h = S$, for rectangular prismoids, or equivalent solids; or, when triangular prismoids are under computation, $\dfrac{2b + 2t + 8m}{12} \times h = S$, equivalent in using triangular sections and double areas, to this rule in words: *The separate products of hights by widths at each end, plus product of sums of hights and widths at both ends, and the sum of these three products, multiplied by $\frac{1}{12} h = Solidity.$*

The following modification of this rule may be sometimes useful in computing the volume of triangular earthwork solids: *The products of the direct multiplication of hight by width at each end, plus sum of half products of the cross multiplications of alternate hights and widths a* *both ends, multiplied by $\frac{1}{6} h$ = solidity from ground to intersection of slopes, and minus the grade prism = solidity from road-bed to ground.*

Many other expressions are assumed for special purposes by the *Prismoidal Formula;* but no matter into what shape it be transformed, the essential idea must always be borne in mind that this formula, *in words, concisely is,*

"The sum of the areas of the two ends, and four times the section in the middle, multiplied into $\frac{1}{6} h$ = S." (*Hutton,* 1770.)

Such is the simple expression of this celebrated formula—given a century ago—which applies not only to all prismoids, but to all right-lined solids, and many curved ones too.*

SUMMARY.

Article.	Formula.	
		For rectangular prismoids, or *any* prismoid, reduced to an equivalent rectangular section, we have Simpson's original rule expressed by sides of the end rectangles, referring to *Fig.* 2, *Art.* **2.** But it is more convenient, perhaps, for our purpose, to designate these sides relatively, *as hights and widths,* and in this form we may write Simpson's rule as follows:
2.	**I.**	(Hight $\times$ Width of one end) + (Hight $\times$ Width of other end) + (Sum of Hights $\times$ Sum of Widths of both ends) $\times \frac{1}{6} h$ = S.
		And the transformation of this formula, for use in the computation of triangular prismoids (*like earthwork*), placing it in Hutton's form.
2.	**II.**	$\dfrac{2b + 2t + 8m}{12}$ = Pris. Mean Area, and $\times h$ = *Solidity.*
		For rectangular prismoids, considered as two wedges.
3.	**III.**	We have Hutton's *General Rule* for any prismoid, $$\dfrac{(b + t + 4m) \times h}{6} = S.$$
3.	**IV.**	We have also Hutton's *Particular Rule.* $$(\overline{2L + l} \times B + \overline{2l + L} \times b) \times \tfrac{1}{6} h = S.$$

* The English engineers have for many years unhesitatingly applied this formula to the warped solids of earthwork. See *Dempsey's* Practical Railway Engineer, 4th edition, 4to, London (1855), pp. 71 to 74. And in this country, Prof. Gillespie (1857), and John Warner, A. M. (1861), have also discussed *the subject of Warped Solids of Earthwork.*

SUMMARY—*Continued.*

Article.	Formula.	
3.	**V.**	For unusual and irregular prismoids we have the method of "*Initial Prismoids*," deduced from Hutton.
6.	**VI.**	For a prismoid, composed of a prism and wedge, superposed.

$$\frac{(B + b + b) \times (H - h)}{6} + (h^2 r - \text{grade triangle}) \times$$

$$h = S.$$

7.	**VII.**	For a trapezoidal prismoid of earthwork, taken as two wedges.

We have the following Rule:

In 1st cross-section $\Big\{$ Add road-bed + top-width + road-bed of 2d section; multiply the sum of these three by level hight of section, and reserve the product.

In 2d cross-section $\Big\{$ Add road-bed + top-width + top-width of 1st section; multiply the sum of these three by level hight of section, and reserve the product.

Finally, add the two products reserved, and $\frac{1}{6}$ of their sum is the mean area of the Prismoid, which, multiplied by length = *Solidity.*

For a triangular prismoid of earthwork, we have the following modification of the Prismoidal Formula, operating by direct and cross-multiplication of hights and widths. All hights being taken at centre from ground to intersection of slopes, and all widths from top to top of slopes on both sides of centre.

Let h and h' = the hights. w and w' = the widths.

Then,

9.	**VIII.**	$\left\{\begin{array}{c}\text{Hights. Widths.} \\ h \times w \\ \times \\ h' \times w' \\ \text{Length} = 100, \\ \text{usually.}\end{array}\right\}$, and $\dfrac{hw + h'w' + \dfrac{hw' + h'w}{2}}{6} \times$ length = S.

SUMMARY—*Continued*

Article.	Formula.	

Simpson's Rule, for the Quadrature and Cubature of Curves (adopted by Hutton), and copied from the 4to Mens. (1770).

10. **IX.**

$$\begin{cases} \text{Sum extreme ordinates} = \text{A.} \\ \quad\text{"} \quad \text{all even} \quad\text{"} \quad = \text{B.} \\ \quad\text{"} \quad \text{all odd} \quad\text{"} \quad = \text{C.} \\ \text{Common distance} = \text{D.} \end{cases} \begin{cases} \dfrac{A + 4B + 2C}{3} \times \\ D = \textit{area or solidity.} \end{cases}$$

For convenience we may transform this into,

10. **X.**

$$\frac{A + 4B + 2C}{6} \times 2D = \textit{area or solidity.}$$

To find the solidity of a triangular prismoid *by roots and squares.*

h and h' = The end hights or representative square roots of the areas of the ends (between ground and intersection of slopes), at regular stations, numbered *even.*

m = Place of mid-section, represented by its ordinate, and numbered *odd.*

Length = Usually, 100, between principal stations.

10. **XI.**

$$\frac{h^2 + h'^2 + (h + h')^2}{6} \times \text{length} = S.$$

Which, for *one station*, is equivalent to Hutton's rule above. This is a very important transformation of the *Prismoidal Formula*, and should be well considered, with the examples in *Art.* **10.**

One of the earliest followers, in the path projected by Sir John Macneill, of using the Prismoidal Formula, with auxiliary tables, for correctly computing the volume of earthwork solids, was G. P. Bidder, C. E., who adopted the obvious plan of imagining the side-slopes to be moved parallel inward, *to intersect at grade,* and then computing the triangular solid thus formed as a prismoid, or the frustum of a pyramid (*both being equivalent in these circumstances*); finally, calculating the centre part (*or core*) as a prism separately, and adding the two for the volume of the whole. The core being computed for one foot wide only,

Article.	Formula.

SUMMARY—Continued.

and then multiplied by the width of road-bed intended to be given.* (This is the plan of Macneill's second series of Tables, for various side-slopes, and base of one foot.)

Bidder's formula for the slopes united is, $[\,(a + b)^2 - a\,b\,]\; \frac{22}{27} = S$, in cubic yards for a 66 foot chain, a and b being the hights or depths at the ends.

This is identical with the formulas of Baker, Bashforth, and others, of subsequent writers: $= (a^2 + a\,b + b^2)\; \frac{22}{27} = S$, in cubic yards, *and is in fact* the algebraic expression for the volume of the frustum of a triangular pyramid, demonstrated in all the elements of geometry—supposed to have been originated by Euclid (about 300 B.C.), and known in this country *as the method of Geometrical Average.*

These formulas are *equivalent* to the following, mentioned in *Art.* **12.**

12.	**XII.**

$$\frac{(\text{Sum of sqs. of hts.}) + (\text{Sq. of sum of hts.})}{6} \times h = S$$

$$= \frac{2\,(\text{Sum sqs.}) + 2\,(\text{Rect. of hights})}{6}, \text{ or dividing by 2,}$$

$$= \frac{(\text{Sum sqs. of hights}) + (\text{Rect. of hights})}{3} \times h = S,$$

which, for a four pole chain, and cubic yards, becomes equivalent to the formulas above, by introducing the proper fractional multipliers—the hights are the square roots of the areas.

---

* A similar plan of computing and tabulating the slopes and core separately: the latter on a base of *unity*, to be subsequently multiplied, by any road-bed, is also that of E. F. Johnson, C. E.—the pioneer of Earthwork Tables in this country (New York, 1840)—and has been followed by several other writers; indeed, it is a method so obvious as to be likely to occur to any student. This *core and slope method* originated by Bidder and Johnson (some 30 years ago), and since repeated by numerous writers, is now again reiterated by the latest compiler of Earthwork Tables, E. C. Rice, C. E. (St. Louis, Mo., 1870).

CHAPTER II.

17...... Since 1833—the date of publication of Sir John Macneill's meritorious volume on the mensuration of earthworks, for canals, roads, and railroads—the investigations of numerous able writers in various countries have shown, *conclusively*, that the Prismoidal Formula (adopted by Macneill) furnishes *the most convenient, if not the only correct rule* for the measurement of the immense bodies of material employed in earthworks, and removed *from*, or supplied *to*, the irregularities of the ground encountered by the location of lines, *under the general name of excavation or embankment.*

The writer, as long ago as 1840, in the Journal of the Franklin Institute of Pennsylvania, repeated the demonstration of the formula referred to, *by means of a simple figure*, and established its connection with the ordinary rules for the volume of the three principal right-lined bodies, known to solid mensuration—*the Prism, Wedge, and Pyramid*—(to all of which, whether complete or truncated, the Prismoidal Formula correctly applies); these are the elementary solids which enter into the composition of a station of earthwork, *and separately, or together, are all computable by the same rule.*

He also showed, by numerous examples (worked out in detail) of the leading forms assumed by railroad earthworks, that by means of *hypothetical* mid-sections, *deduced* from the usual cross-sections taken in the field (and diagrammed between them if necessary), the volumes of excavation and embankment solids could be computed correctly without unusual labor, *and with more than usual accuracy.* This method was made to depend essentially upon two points : *

---

* Journal of the Franklin Institute (Philadelphia, 1840).

1. "That the formula expressing the capacity of a prismoid *is the fundamental rule* for the mensuration of all right-lined solids, whose terminations lie in parallel planes, and is equally applicable to each."

2. "That any solid whatever, bounded *by planes*, and parallel ends, may be regarded as composed of some combination of prisms, prismoids, pyramids, and wedges, or their frusta, having *a common altitude*, and hence capable of computation by the general rule for prismoids."

All excavation and embankment solids come within the scope of these definitions, and *all* are computable with ease and accuracy by means of the Prismoidal Formula.

These views have met with general acceptance from most practical writers, but many useful transformations and modifications have naturally been indicated; all grounded upon the same formula which appears to have originated with THOMAS SIMPSON, an eminent mathematician, and was demonstrated and published by him (*for rectangular prismoids*) in London, 1750 (*Arts.* **1** and **2**), but generalized and made more useful by HUTTON, in 1770 (*Art.* **3**).

This extraordinary formula is not only the fundamental rule for all right-lined solids, but reaches also to many curved bodies and warped surfaces (as before mentioned), so that it may safely be assumed *as correct* for all the earthwork solids in common use, which, indeed, are invariably laid out with the view of reducing the ground, *however irregular*, to equivalent planes (*as near as may be*), by means of levels and sections, taken at short distances; and though this effort may not be entirely successful in practice, it must be so nearly so that the warped surfaces, remaining involved in the solid, can only differ slightly (*if at all*) from those for which the Prismoidal Formula is known to hold.

As *a general rule*, it may therefore be considered as close an approximation to existing facts as is admitted by any convenient method within the present range of human knowledge, and far more accurate than *any* of the *proximate* rules, which have been extensively employed for the solution of the complicated problems of earthwork.

As a preliminary matter, it is necessary now to make some remarks on the manner of collecting data in the field, for subsequent use in calculating the quantities of earthwork solids.

The centre or guiding line of the road or work having been carefully located upon the ground, and marked off in regular stations—

usually of one hundred feet each—the next operation is to cross-section the work, with *level, rod,* and *tape;* most engineers also using the clinometer, or slope level, as an auxiliary, in some stages of the process. The centre line is assumed in all cases to be *straight,* from point to point, and generally to be a tangent line, to which the cross-sections are perpendicular, but owing to the convergence of the radii upon curves, this is not strictly correct—though within the limits of the work staked out, that convergence is but slight; nevertheless, the cross-sections (before proceeding to level them) should be set out *approximately,* normal to the tangents, and radial to the curves; and upon all curves, or at least on all of small radius, *intermediates at half distance* should be placed, or, if the curves are unusually sharp, *even at the quarter of a regular station.*

Some engineer manuals furnish formula for the correction of quantities upon curved lines,* but they are rarely used; a simple reduction of distance between the cross-sections, or a closer assemblage of them, *being usually deemed sufficient.*

The surface of the ground † is regarded by the engineer as being composed of *planes* variously disposed, with relation to each other, so

---

* The simplest and most convenient rule for this purpose, is that of Warner's Earthwork (1861). This rule has been adopted, and somewhat simplified, by Prof. Rankine, in Useful Rules, etc. (London, 1866).

The process is : First, to calculate the solidity of the earthwork to the intersection of the slopes (as though the line were straight), and then to multiply it by a factor, which corrects for curvature.

This factor is found *thus :* $\dfrac{\text{Difference slope distances}}{3 \text{ Radius of curve.}} \pm 1.$ *The corrective quotient* being added to unity, when the greater slope distance lies outward *from* the curve, or subtracted, *if otherwise.*

For example, take a curve of 700 feet radius, lying upon a heavy embankment, along a ground surface sloping uniformly inwards, *towards* the centre of the curve, at the rate of 15°. The road-bed being 24 feet wide, and side-slopes $1\frac{1}{2}$ to 1.

Let the difference of slope distances be 42 feet, the greater slope distance *inwards,* and suppose the whole volume, for straight work = 5917 cubic yards to intersection of slope. *Then,* $\dfrac{42}{3 \times 700} = \cdot 02,$ and $1 - \cdot 02 = \cdot 98,$ *the factor required.* Then, $5917 \times \cdot 98 = 5799$ cubic yards, and $5799 -$ grade prism (356) $= 5443$ cubic yards, *the volume, corrected for curvature.* The difference in this case, produced by the curvature of the line, being 118 cubic yards, for the station computed.

The correction for other curves would be *inversely* as their radii, and for a 1° curve, similarly situated, about 15 cubic yards, *per station.*

The difference of the *distances out* from the centre are the same thing as Prof. Rankine's difference of slope distances—since the former involve an equivalent quantity on both sides of centre, equal to half the road-bed.

† Journal Franklin Institute (1840).

6

that any vertical section will exhibit a rectilineal figure, more or less regular. This supposition, though not strictly correct, is sufficiently accurate for practical purposes.

Upon the cross-sections (taken near enough together to define positively the general figure of the surface), sufficient level points are obtained transversely, by *level and rod*, their distances out from centre being simultaneously measured, with *a tape line;* in this manner, both vertically and horizontally, in relation to established planes, the position of all the points necessary to determine the configuration of the ground is well ascertained.

These points of elevation, or depression, are commonly called *plus* or *minus* cuttings (or simply *cuttings*), and the horizontal distances which fix their relation to the centre are shortly called *distances out.*

The details of the operation of *taking the cuttings, or cross-sectioning the work* (a matter of vital importance in correct measurement), require good judgment and accuracy; but are so well known to practical engineers as to render unnecessary a description *at length.* This operation, however, is the absolute foundation upon which the whole fabric of computation rests, and if it be not *judiciously executed,* all rules are vain.

We may here mention a general maxim, which should never be neglected, if accurate results are desired, viz.: *At every change of surface slope, transversely, single cuttings and distances out must be taken; and at every longitudinal change, sections of cuttings, or cross-sections.*

Upon very rough ground it is customary to make the lateral distances apart of the cuttings, uniformly 10 feet, which materially facilitates the subsequent calculations; so much so, indeed, that on a rock side hill it is often advisable to use this distance, even though the ground seems not actually to need it; the cuttings and distances out are commonly taken in feet and tenths, and the regular stations of one hundred feet are subdivided by cross-sections into shorter lengths, if the ground requires it, *as is frequently the case.* *One foot* being usually the unit of linear measure, *one hundred feet* a regular station, and *the cubic yard* the unit of solidity, in earthwork.

Though not indispensably necessary, it will be found convenient in using the prismoidal method of calculation, as well as conducive both to expedition and accuracy, to observe the following rules in "*taking the cuttings,*" as far as the character of the surface will admit, viz.:

1. On side-hill, at each cross-section, where the work runs partly in filling and partly in cutting, ascertain the point where grade, or bottom, strikes ground surface.

2. On every cross-section, take a cutting at both edges of the road, or at the distance out right and left of one-half the base.

3. Always take a cross-section, whenever either edge of the road-bed strikes ground surface, and set a grade peg there to guide the workmen.

4. On rough side-hill, or wherever the ground appears to require it, take the cuttings (not otherwise provided for) at ten feet apart.

5. Wherever the ground admits, place the cross-sections at some decimal division of 100 feet apart, as 10, 20, 30, etc.

6. Endeavor to take the same number of cuttings, in each adjacent cross-section, to facilitate the computation.

7. On plain and regular ground, take three cuttings only—at centre and both slopes.

If these simple directions are observed by the field engineer, and the work carefully done, much labor will be saved, both to him, and to the computer in the office.

In all cases of side-long ground, we suppose it to slope in the same general direction, between the end sections, and do not admit of *opposite* surface slopes, because, under the general rule, the field engineer would place a cross-section at the point of change slope, and render the consideration of opposite slopes, and the warped surfaces they always produce, *entirely unnecessary;* indeed, by more closely assembling the cross-sections together, we can practically *reduce* even the most irregular surface to a series of planes coincident with it.

Nevertheless, an able writer * has shown that warped solids of a certain kind are computable by *his* rules; and the late Professor Gillespie, in several valuable essays, has demonstrated that hyperbolic paraboloids *at least* could be correctly calculated by the Prismoidal Formula; while English engineers have long used this rule for computing the volume of earthwork solids, *with warped surfaces;* † it appears, however, to be more certain and satisfactory if we confine the operations of this formula *to solids bounded by plane surfaces* as nearly as circumstances admit; but it is fortunate that our rule is

---

* John Warner, A. M., Computation of Earthwork (1861).—Prof. Gillespie, Manual of Roads and Railroads, 10th edition (1871).

† Dempsey, Practical Railway Engineer (London, 1855).

known to hold for *some* descriptions of warped ground, and hence can hardly fail to proximate results, near unto the truth, however much the surface may be warped, between the cross-sections, if they have been judiciously placed by the field engineer.

a. ...... The modification of the Prismoidal Formula, which we shall employ in this first method of computation, will be that designed to find *a mean area*, to be subsequently employed by the aid of our Table, at the end, to ascertain the cubic yards of volume.

This formula comes from that generalized by Hutton (1770) through the special mid-section, and is expressed in the beginning of *Art.* **16** as follows : *

$$\frac{b + t + 4m}{6} = Prismoidal\ Mean,\ \text{and} \times h = \text{S (the Solidity).}$$

Summarily expressed in words *as follows;* One-sixth the sum of end areas, and quadruple mid-section, multiplied by length, gives the *Solidity.*

This general formula (identical with one of Hutton's) requires three areas (one, the mid-section, deduced from the others), and also the hight or length of the Prismoid *to be given;* and by its aid we propose in illustration to furnish *five* examples of calculation.

    1. Of a regular station, of *three-level* ground.
    2. Of the same length, of *five-level* ground.
    3. Of *seven-level* ground.
    4. Of *nine-level* ground.
    5. Of a portion of excavation and of embankment *adjacent,* with an oblique passage between them, from one to the other.

We here follow a classification of ground nearly resembling that adopted by the late Prof. Gillespie (one of our ablest writers upon earthwork), who enumerates four classes only, under the simple nomenclature of, 1, *one-level;* 2, *two-level;* 3, *three-level;* 4, *irregular ground;* and under these four classes, he dealt with the problems of earthwork in his excellent lectures "to the Civil Engineering Classes in Union College." †

---

   * "This rule," says Prof. Rankine, in Useful Rules and Tables, 2d edition, London, 1867, p. 74, " applies generally to any solid bounded endwise by a pair of parallel planes, and sideways by a conical, spherical, or ellipsoidal surface, or by any number of planes."

   † Manual of Roads and Railroads, 10th edition (1871).

We think, however, that few engineers would be willing to class ordinary *five-level ground* as *irregular;* for such ground would in fact be produced simply by the angle levels commonly taken, which at once convert the plainest *three-level* into *five-level ground.*

But ground requiring *more* than five cuttings *on one cross-section,* all would probably agree in classifying as *irregular,* and such is the view taken by the present writer.

This would bring all ground whatever within the scope of *five classes,* and make but a slight variation in Gillespie's nomenclature. 1. Level ground, where the centre cutting alone is sufficient for volume. 2. Ground slightly inclined, where side-hights only may have been taken. 3. Ordinary ground, requiring centre and side-hights. 4. Same as 3, with the addition of angle levels, or one cutting right and left of centre, besides those at the slope stakes. 5. Irregular ground,—such, or any similar classification would somewhat simplify the matter of earthwork, but it is not *indispensable.* Centre cuttings, or level hights at the centre, are, however, invariably taken in the field, and recorded at the time, whether they be subsequently used or not, so that class 2 would seldom occur on original ground.

The method of measuring the capacity of long irregular solids, by means of normal sections, at short distances, has long been used by mathematicians; of which numerous examples may be found in Hutton (1770), as well as in the demonstration and use of Simpson's rule for quadrature and cubature, referred to in many works, both civil and military.

This method then was naturally adopted by the earlier engineers for the mensuration of earthwork, and has been continued down to the present day with little chance of being superseded; as the areas of the sections, commonly known to the engineer as *cross-sections,* are not only useful in the computation of solidity, but also in many other ways, during the progress of earthworks; and consequently those rules which disregard the areas of cross-sections, and aim directly at the volume alone of excavation and embankment, *are less useful (even if more concise) than those which require the sectional areas to be first computed.*

18. *Examples in Computation by the First Method.*

In computing by this method, the Grade Prism is **not required**, *and is not used*, but it may be employed in verification.

Example 1.— We will now give three figures (*Figs.* 53, 54, and 55), representing **three** cross-sections, upon one regular station of 100 feet in length, of a railroad cut with side-slopes of 1 to 1, and road-bed of 20 feet—the other dimensions being as marked upon the figures.

In these, the first and last represent the end cross-sections of the 100 feet station, supposed to have been regularly taken in the field.

The other (*Fig.* 54) being the *hypothetical mid-section*, deduced from the end ones, as required by HUTTON's General Rule.

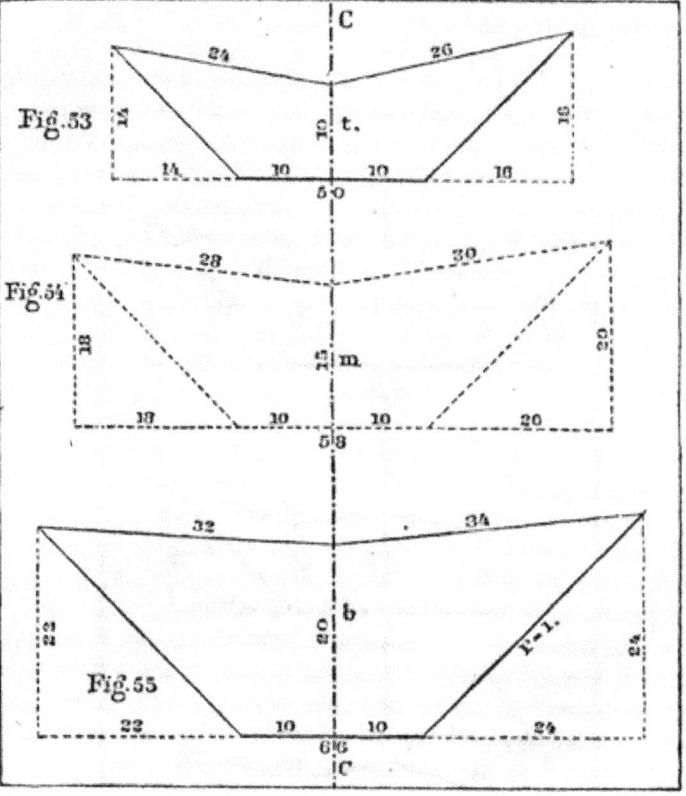

These cross-sections are marked *as follows*:

$$\left\{\begin{array}{l} b \ = \ 890 \text{ Area.} \\ m \ = \ 625 \quad " \\ t \ = \ 400 \quad " \\ \text{Length, 100 feet} = h. \end{array}\right\} \textit{Example 1.}$$

And the calculations for *solidity* are as below:

$$\text{Calculations,}\begin{cases} \quad 890 \quad = b. \\ \quad 400 \quad = t. \\ \quad 2500 \quad = 4\,m. \\ \overline{6\,)3790} \\ \quad \overline{631\cdot7} \; = \text{Prismoidal Mean Area.} \\ \quad \overline{2339\cdot6} \; = \text{Cubic Yards (by Table) for 100 feet.} \end{cases}$$

The above example is for plain ground of "*three levels,*" as classed by Professor Gillespie.

Example 2.—We will now give an example of a railroad cut, with the same road-bed (20) and ratio of side-slopes (1 to 1), in *five-level ground.*

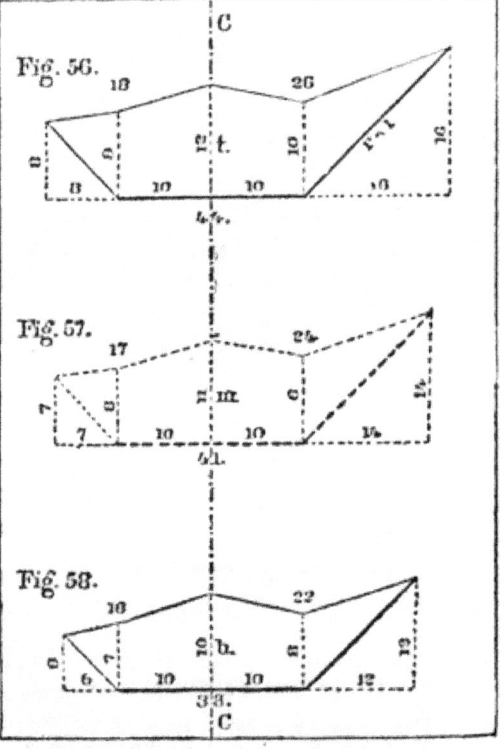

The three cross-sections, upon the regular station of 100 feet, are numbered, *Figs.* 56, 57, and 58, and marked *b*, *m*, and *t*, the middle one being Hutton's *hypothetical* mid-section, deduced by Arithmetical Averages from *b* and *t*, the cross-sections, assumed to have been taken in the field, with *rod, level,* and *tape,* in the usual manner.

$$
Example\ 2 \left\{ \begin{array}{l}
\text{Cross-sections.} \\
b = 244\ \text{Area.} \\
m = 286\ \text{``} \\
t = 331\ \text{``} \\
\text{Length 100 feet} = h.
\end{array} \right\}
$$

And the calculations for *solidity* are as follows:

$$
\begin{array}{ll}
244 & = b. \\
1144 & = 4m. \\
\underline{331} & = t. \\
6)\overline{1719} \\
\hline
286\cdot5 & = \text{Prismoidal Mean Area.}
\end{array}
$$

And for Cubic Yards, in 100 feet long, per Table = 1061·1.

Example 3.—We will now give an example of a railroad cut, similar to the preceding, base 20, slope ratio $r = 1$, *in seven-level ground.*

$$
Example\ 3 \left\{ \begin{array}{l}
\text{Cross-sections and areas.} \\
b = 524 \\
m = 537 \\
t = 551 \\
\text{Length, 100 feet} = h.
\end{array} \right\}
$$

Calculations for solidity:

$$
\begin{array}{ll}
524 & = b. \\
2148 & = 4\ m. \\
\underline{551} & = t. \\
6)\overline{3223} \\
\hline
537\cdot2 & = \text{Prismoidal Mean Area.}
\end{array}
$$

And for Cubic Yards, in 100 feet long, per Table = 1989·6.

Example 4.—Although embankment is merely excavation *inverted,* and governed in its computation by precisely the same principles, we will now give an example of embankment on irregular *or nine-level ground,* road-bed 16, side-slopes 1½ to 1, and ground surface supposed to be jagged masses of rock. CC represents as usual the centre or guiding line of the road, the cross-sections being *dimensioned* *re*

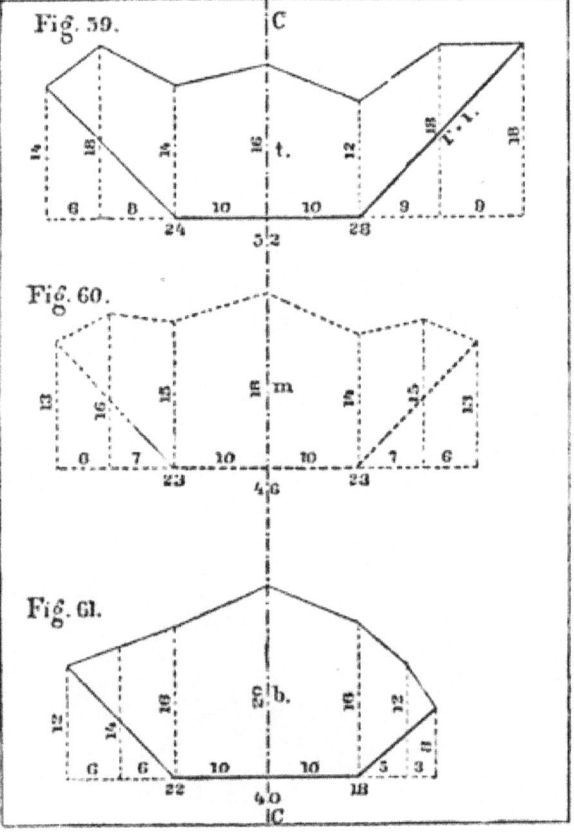

Fig. 59.

Fig. 60.

Fig. 61.

marked upon the figures (62, 63, 64), the distance between the end sections being a regular station of 100 feet, and *m* (*Fig.* 63) being the *hypothetical* mid-section, deduced from the two others, supposed to have been regularly measured by the field engineer, and furnished to the computer by him from his note book.

The areas of the sections being *given*, having been previously calculated in the customary manner.

$$
\text{Example 4} \left\{ \begin{array}{l} \text{Cross-sections and areas.} \\ b \ = 602 \\ m \ = 691 \\ t \ = 786 \\ \text{Length, 100 feet} = h. \end{array} \right\}
$$

Calculations for solidity;

$$602 = b.$$
$$2764 = 4\,m.$$
$$786 = t.$$
$$6)4152$$
$$692 = \text{Prismoidal Mean Area.}$$

And for Cubic Yards, in 100 feet long, per Table $= 2562{\cdot}9.$

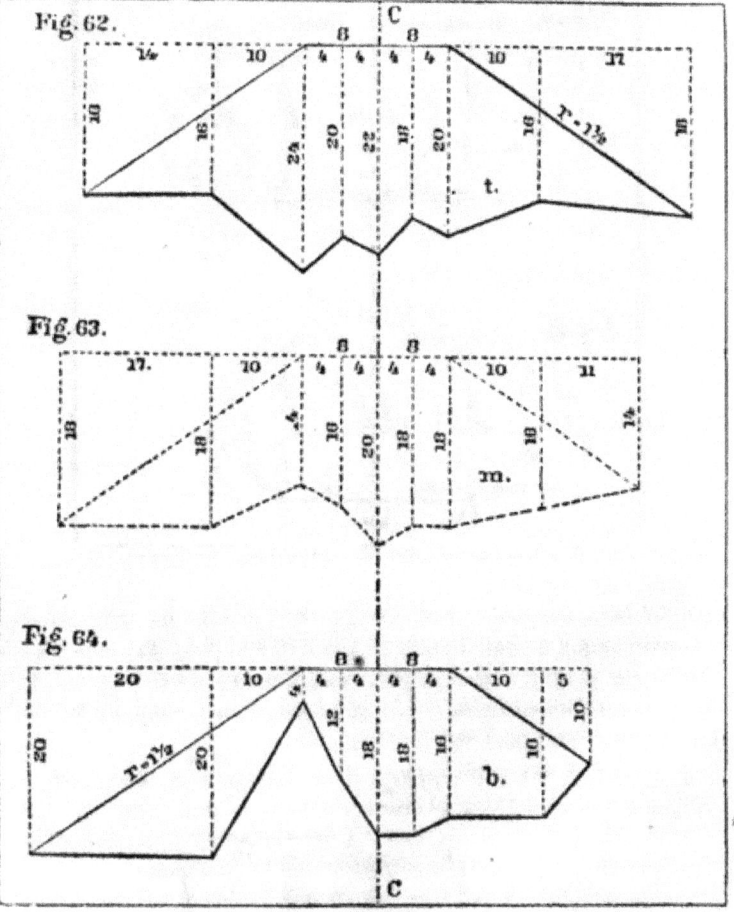

As has been observed before, b and t are correlative, and either might be taken as base; the calculations of quantity are usually made in the direction in which the numbers run, or the one nearest to us of any pair may be assumed as b, and the other as t—it is quite immaterial which—but during the pendency of the computation, to which they are subject, the special designation must remain for the time *unchanged*.

The surface of ground, assumed in this example, appears to be *sufficiently irregular* to test any rule (though rougher ones will occur to the memory of most engineers), and we might proceed to give illustrations of such, but enough has been done in this way to indicate the principles on which we work, and which can readily be applied to any case which may occur in practice. Nor does it seem necessary here to define and classify the numerous distinct cases of earthwork— *the Prismoidal Formula holds for all*, and it is left to the judgment of the engineer to make the application.

19. *Connected Calculation of Contiguous Portions of Excavation and Embankment, with the Passage from one to the other.*

Example 5.—See *Figs.* 65 to 71.

In *Fig.* 65, ABC, a portion of a railroad *cut*, road-bed = 20, side-slopes 1 to 1. BCD, a portion of a railroad *fill*, road-bed = 14, slopes 1½ to 1. Grade points ☉ four in number, besides the centre.

In *Figs.* 66 to 71, six cross-sections, 3 of excavation and 3 of embankment, are shown, and all *dimensioned* as marked. *Fig.* 68 is the base of the closing pyramid of excavation in the passage from excavation and embankment, the vertex of which is at the grade point B. *Fig.* 69 is the base of the closing pyramid of embankment, in the passage from embankment to excavation, the vertex of which is at the grade point C.

The other cross-sections are those necessary to compute the portions of excavation and embankment shown upon the plan, *Fig.* 65. One of them only is at a regular station, called station (10), *Fig.* 68, the others are all *intermediates*, supposed to have been required by the configuration of the ground.

The scale is 20 feet to the inch.

On the centre line, the excavation shown is 61 feet in length—but the closing pyramid of cutting runs 11 feet further to its vertex at the grade point B. While in like manner the embankment is 48 feet long on the centre, and the closing pyramid of filling extends 7 feet further to its vertex at the grade point C.

This over-lapping of the closing pyramids is an inconvenience, but it is sometimes *unavoidable*.

Plan

Cross Secs.

Fig. 65.

Fig. 66.

Fig. 67.

Fig. 68.

Fig. 69.

Fig. 70.

Fig. 71.

Calculations for Solidity.

Position of Cross-sections upon the centre.	Distances apart.	Cross-section Areas, etc.	
9 + 50 . . .	0 . . .	342 = b.	
9 + 75 . . .	25 . . .	907 = 4 m.	
10 Reg. Sta. . .	25 . . .	106 = t.	} Excavation.
Length = 50		6)1355	
		225·8 = Prism. Mean Area.	
		418·1 = Cubic Yards, by	
		Table for $\frac{50}{100}$ feet = 418·1	

10 + 11 Grade at centre.

(*Passage*, etc., from *Excavation* to *Embankment*.)

Closing Pyramid of Excavation, vertex at B, *Fig.* 65.

Area of base at 10 = 106. Then,

$$\frac{106 + 106 + 0}{6} \overset{\text{Mean Area.}}{=} 35\cdot3 \times \text{length}, 22 = \text{by Table } 130\cdot7 \times \tfrac{22}{100} = \quad 28\cdot8$$

Total *Solidity* of Excavation . . . . = 446·9

Now, commence the embankment with the closing pyramid in the passage, altitude or length 15 feet, and vertex at C, *Fig.* 65. Area of base at 10 + 19 = 46. Then,

$$\frac{46 + 46 + 0}{6} \overset{\text{Mean Area.}}{=} 15\cdot3 \times \text{length}, 15 = \text{by Table } 56\cdot7 \times \tfrac{15}{100} = \quad 8\cdot5$$

10 + 19 . . .	0 . . .	46 = b.	
10 + 39 . . .	20 . . .	504 = 4 m.	
10 + 59 . . .	20 . . .	215·5 = t.	} Embankment.
Length = 40		6)765·5	
		127·6 = Pris. Mean Area.	
		189·0 = Cubic Yards, by	
		Table for $\frac{40}{100}$ = 189·0	

Total *Solidity* of Embankment . . . . . . = 197·5

And this closes the computation of Cubic Yards in the portion of Excavation and Embankment, from A to D (*Fig.* 65), including the passage between them, *and comprising in all two prismoids and two closing pyramids.*

In concluding this branch of the subject, we may mention that as HUTTON defines "*a prismoid*" to have in its end sections "*an equal number of sides*" (*Arts.* 3 and 14), a like number of level hights, or cuttings, ought always to be taken in adjacent cross-sections, but should that have been omitted in the field, additional cuttings may be computed or drawn upon the sections obtained, so that previous to calculating their areas, *there shall be the same number of cuttings in all the adjacent cross-sections, and we shall then have for solidity a correct prismoid.*

a. . . . . . . In verifying the work given in the first four examples preceding—illustrated by *Figs.* 53 to 63 inclusive—the end areas and length being correctly given in all, it is only necessary to *prove* the mid-section ; as an agreement there necessitates a like result when used with the given data, *prismoidally*, to find the solidity.

This proof may be made either by our 2d method of computation (Hights and Widths), or 3d method (Roots and Squares)—the latter being generally the most convenient, though the former may often be used with advantage.

No *single* calculation, truly says Prof. Gillespie, ought ever to be relied on by the engineer, and proof of the correctness of every computation should always be obtained before employing it in work.

It is often the case when railroads follow the rugged margins of rivers that many miles of side-hill work present themselves, where the road-bed, located above the flood line, lays in rock excavation on one side, and heavy embankment upon the other—to such cases the preceding method of computation will be found peculiarly applicable ; both cutting and filling showing themselves upon the end cross-sections of every station and intermediate, while the mid-section may be *diagrammed* between them with great facility.

In continuing this chapter we may state—*That in any right-lined solid whatever*, lying between two parallel planes (according to the definition of a prismoid), whenever a mid-section can be correctly deduced between two given end sections, situated in the limiting planes (and by taking pains it always can be), there, our First Method of Computation *will be found to apply strictly for solidity.*

So that this method is a standard test for all other rules, and has been accepted as such by Prof. Gillespie, and other able writers.

Hence, we may repeat that the formula employed in this chapter *is the fundamental rule for the mensuration of all right-lined solids, within parallel planes,* and applicable also to many warped figures, and other curvilinear bodies, in a manner so unexpected as to have excited the surprise of some able geometers, whose attention had not been specially directed to that subject before.

Cases often occur in heavy work, where it is evident from the cross-sections, that the bulk of the solid under consideration lays considerably on one side of the centre line (or where, in common phrase, the sections are *lop-sided*), and it would seem in such cases as if some correction ought to be made for the position of the centres of gravity (as indicated upon *Figs.* 43 and 44, Chapter I.); for it is most obvious that in a long line of heavy work the path of gravity centres would frequently *cross and re-cross* the guiding line of the work, and hence *would necessarily be longer.*

So that if the line of magnitude should be assumed as the true line of calculation, the centres of gravity ought to be assembled upon the centre line, *in effect,* at every station, and this correction would probably be found by multiplying the projections of the points of gravity upon the centre, by their distances from it (+ when on the same side — when opposite); but this is a refinement which has never been employed by engineers, in dealing with the huge masses in question.

What the engineer most needs in earthworks appears to be—not astronomical accuracy, but *the systematic use* of some rule for solidity, which shall always be consistent with itself, and closely proximate the truth, without involving those stupendous discrepancies (mentioned by many writers), as flowing from the employment of the *average* methods, which have been so much (and as it always appeared to the writer) so unnecessarily, *used in the ordinary computations of earthwork.*

The method of computation developed in this chapter finds appropriate application also *in masonry calculations.* In this manner the writer once computed the contents of a heavy stone aqueduct, containing over 4000 perches, with numerous projections and off-sets, and walls battered, *both inside and outside.*

The process taken was by drawing to a scale accurate horizontal plans, at all the off-set levels, at the skewbacks, and other breaks in the contour—deducing mid-sections between these, and multiplying together each set of three, in accordance with the Prismoidal Formula, etc.

This gave a very satisfactory exhibit of the work, and a correct result *in volume,* with less labor, and greater accuracy, than any other modes he found in use at the time.

In calculating stone culverts, and bridge abutments also, this method will be found quite useful.

In fact, in computing the volume of solid bodies of any kind, the engineer will find the Prismoidal Formula *to be either strictly correct, or a very close approximation.*

b. . . . . . . We now conclude this chapter by some remarks upon *Borden's Problem.*

Some examples acquire celebrity from being apposite in themselves, for the illustration of important processes, and are consequently copied by others; besides, there is an evident advantage to the reader in re-producing examples, which, having been before discussed, are more generally known; amongst such is *Borden's Problem,* first published by Simeon Borden, C. E. (Boston, 1851), in his "System of Useful Formulæ" (*Art.* 63).

He treats this example at great length (14 pages), and commits some errors, which were subsequently pointed out and corrected in Henck's Field Book (Boston, 1854).

This example was also adopted by John Warner, A. M., in his Earthwork (Philadelphia, 1861, *Art.* 112), without comment.

The problem appears to have given Mr. Borden some trouble, involving a number of his "*blind pyramids,*" and also some errors, as Mr. Henck hath shown.

Nevertheless, it is simply a case of *injudicious cross-sectioning*—for had Borden, instead of attempting to compute its full length of 100 feet, imagined an intermediate at 50 feet (for which he gave all the data necessary), all difficulty would have vanished, and he would neither have stumbled over his own blind pyramids, nor been shortly corrected by a subsequent author.

Indeed, Mr. Borden admits, page 186, of his work of 1851, that "the engineer would be likely to divide the section into two or three" —and this the present writer deems to be not only likely, *but absolutely certain.*

Now, taking the end areas alone (100 feet apart), and disregarding (for the moment) the irregularities of the ground, which ought to have been intercepted and brought out, by an intermediate at 50 feet—we find:

Warner, in *Art.* 112, of his Earthwork, gives for
the volume . . . . = 1155·9 C. Yards.
By Hutton's General Rule (as in this chapter) = 1155·9   "
Difference . . . . = 0

But Henck, in his Engineer's Field Book (after noting Borden's mistake of 360 cubic feet), finds by his own process the solidity =

32,820 cubic feet = 1215·5 cubic yards; or, the former are in a deficiency of — 59·6 cubic yards, an error inadmissible in the quantity before us.

In this problem Borden makes two theoretical suppositions, and two summations of results, based upon his hypothetical view of the effect upon solidity of the irregularities of the ground surface, between the end sections, *but he gives no opinion on either.*

The Prismoidal Formula of Hutton (computed on the whole station of 100 feet) *gives precisely an Arithmetical Mean between the two suppositions of Borden*, but is considerably in defect of the true volume as given by Henck's Formula.

And here we come to the point of the importance of properly cross sectioning a solid, before we begin to calculate it;—for if we sketch from Borden's data *an intermediate* at 50 feet, of which we find the area to be 335·6—*then all difficulties are at once resolved*, and we proceed prismoidally in a few lines to reach *a correct result*, which Mr. Borden failed to attain in fourteen pages.

Considered in connection with an intermediate at 50 feet, *Borden's Problem* stands as follows: Two end areas = 387 and 240. One intermediate area = 335·6. Now, deducing between these (by Borden's data) the hypothetical mid-sections, required by Hutton's General Rule, we find they have areas of 293·5 and 366·5, and working *prismoidally* with them we quickly find *the solidity* of the entire body to be 32,820 cubic feet, or 1215·5 cubic yards—*precisely the same* as Henck makes it by his own formula, and as Borden would have made it had he been aware of the errors into which his own "*blind pyramids*" led him.

As this problem is a well-known one, and has not *a very irregular* appearance in Borden's diagram, we think this a suitable place to urge upon all engineers *the great importance of judicious cross-sectioning.*

In terminating this chapter, we may safely state that Hutton's General Rule, as applied to earthworks by the methods detailed herein, IS ONE WHICH NEVER FAILS WHEN THE DATA IS CORRECT.

7

CHAPTER III.

20. ...... *The Prismoidal Formula*, as originally demonstrated
by Simpson (1750)—see *Art.* **2**—was evidently designed for the rectangular prismoid (*Fig.* 2)—its end areas were obtained by multiplying together *the Hights and Widths;* and four times its mid-section
by multiplying *the sum of the Hights by the sum of the Widths.*

To adapt it more conveniently to the triangular prismoids of Earthworks, with side-slopes drawn to intersect each other, the original
formula of Simpson (1750), reduced to the form subsequently enun-
ciated by Hutton, *as a general rule* (1770), is multiplied by 2, on the
left side *only,* changing its divisor *at the same time.*

Thus,

$$\frac{(b + t + 4\,m) \times h}{6} = S \times 2 = \frac{2\,b + 2\,t + 8\,m}{12} \times h^* = S.$$

This is the same thing, in effect, as the original formula of Simpson (when arranged for a mean area); for if we suppose the rectangular prismoid (*Fig.* 2) cut in half by a plane through the diagonals
of its end areas, FB, etc., so as to convert it into *two triangular prismoids* (each with one right angle), the Hights $\times$ Widths from the
right angle would give *double* the triangular area of each end, while
their sums, multiplied together, would equal 8 times the triangular
mid-section, the divisor becoming $6 \times 2 = 12$.

---

* It would evidently be a much better notation for earthwork to adopt *l instead of h,*
because the greatest extent of an earthwork solid usually lays along the ground (*length-
wise*); but Simpson and Hutton, the fathers of these formulas, have both used *h*—they
dealing generally with prismoids of small dimensions, supposed to stand erect upon a
base (as in *Figs.* 1 and 3), and have been followed by most writers, and necessarily for
the most part also *here;* though we have occasionally used *l* (to avoid confusion), and
this must be taken as correlative with the *h* of Simpson and Hutton, in the cases in
which it has been employed; but some care will be needed to avoid confounding the *h*
indicating the length of the prismoid, with the same letter often used as a symbol for
hight *in cross sections.*

Now, as shown in *Art.* **8, a,** it is an equivalent process to imagine the triangular section, partially revolved, so as to bring the edge of the dihedral angle *downwards*, and to cause its *bisector* (the centre line) to become the perpendicular *hight* (h) of the cross-section, while the extreme breadth to ground edges of side-slopes, horizontally, becomes the *width* (w)—then, by *Art.* **8,*** we have $h \times w = $ *double area of triangular section to intersection of side-slopes.*

This is the position occupied by the triangular areas of the cross-sections of the solids forming the earthworks of railroads, the centre line being the bisector, or *hight* (h), and the sum of the distances out, to the ground edges of the side-slopes of an equivalent triangle, being the *width* (w).

The equivalent triangle is often formed by means of an equalizing line, drawn (for convenience) through the lowest side-hight of the cross-section, so as to form a figure of only three sides, *exactly equivalent* in area to the cross-section of earthwork, which is nearly always more or less *irregular* on the top, and frequently has numerous sides for its ground line ;—the side-slopes, however, remaining generally uniform and even, from station to station (see *Fig.* 14).

The equation for Hights and Widths may often take another form (already mentioned in *Art.* **9**), *which, at times, will be found convenient.*

$$\text{Let } \begin{cases} h & = \text{ Hight at one end.} \\ h' & = \quad\text{``} \quad \text{`` other end.} \\ w & = \text{ Width at one end.} \\ w' & = \quad\text{``} \quad \text{`` other end.} \\ l & = \text{ Length of mass, usually} \\ & \quad\text{ denoted by } (h) = \\ & \quad 100, \text{ generally.} \end{cases}$$

$$\text{Then, } \frac{h\,w + h'\,w' + \dfrac{h\,w' + h'\,w}{2}}{6} \times l = \text{S.}$$

---

* In any $\triangle$, however situated :—If one angle coincides with the intersection (or origin,) of two rectangular axes (such as a Meridian, and an East and West line, or centre line, and base of levels), and the co-ordinates of the other angles are known (as by their Lat. and Dep., or level hights and distances out); then, *the area* of any such $\triangle$ is easily found.

Thus, calling the first angle 0, and the others in succession 1 and 2.

We have, $\dfrac{(\text{Lat. of } 1 \times \text{Dep. of } 2) - (\text{Lat. of } 2 \times \text{Dep. of } 1)}{2} = \text{Area of } \triangle \text{ required.}$

But, in the single case of either rectangular axis *cutting* the $\triangle$, then, instead of — between the products (forming the numerator above) put +. With this exception, the

This formula may be briefly called (from a leading feature in the process), *the direct and cross multiplication of Hights and Widths*, which may be represented as below; and then, $\left(\times \dfrac{l}{6}\right)$, or *one-sixth the whole being taken = Solidity.*

$$Thus, \quad \dfrac{\left\{\begin{array}{c} h \times w \\ \times \\ h' \times w' \div 2 \end{array}\right. = \left\{\begin{array}{c} h\,w \\ h'\,w' \\ \dfrac{h\,w' + h'\,w}{2} \end{array}\right.}{6} \times l = S.$$

For example, take *Figs.* 72 and 73 (dimensioned as marked).

1. *By Direct and Cross Multiplication of Hights and Widths.*

Direct $\begin{cases} h\,w = 23{\cdot}4 \times 47 & \ldots \ldots = 1100 \text{ Double area.} \\ h'\,w' = 27{\cdot}6 \times 55{\cdot}5 & \ldots \ldots = 1532 \quad " \quad " \end{cases}$

and

Cross Multi- $\begin{cases} h\,w' = 23{\cdot}4 \times 55{\cdot}5 = 1299 \\ h'\,w = 27{\cdot}6 \times 47 \;\; = 1297 \end{cases}$
plication.

$$2)\overline{2596}$$
$$\overline{1298} = 1298 \left\{\begin{array}{l}\text{Representa-} \\ \text{tive \; product} \\ \text{for mid-sec.}\end{array}\right.$$

Let $\begin{cases} h \; = + \; 23{\cdot}4 \\ w \; = \quad 47 \\ h' \; = + \; 27{\cdot}6 \\ w' \; = \quad 55{\cdot}5 \end{cases}$

$$6)\overline{3930}$$

Prism. Mean Area = $655 \left\{\begin{array}{l}\text{Including \; the} \\ \text{grade trian.} \\ \text{of 100 area.}\end{array}\right.$

2. *Proof* by Simpson's Formula (modified for triangles).

$$\left\{\begin{array}{l} \quad \text{Hights.} \quad \text{Widths.} \\ 23{\cdot}4 \times \;\; 47 \;\; = 1100 \\ 27{\cdot}6 \times \;\; 55{\cdot}5 = 1532 \\ \overline{51 \;\; \times 102{\cdot}5 = 5228} \\ \qquad\qquad 12)\overline{7860} \\ Prism.\; Mean\; Area = \;\; 655 \;\; as\; above,\; \text{including} \\ \qquad\qquad\qquad\qquad\qquad\qquad \text{grade triangle.} \end{array}\right\}$$

Then, the *mean area* $\times$ *length* = 100 feet between sections = *Solidity* = 65,500 cubic feet.

---

rule is *general*, and finds ready application *in computing the areas of irregular cross-sections*, and the contents of LAND SURVEYS.

(Prob. V., Young's Analyt. Geom., London, 1833.—Prof. Johnson's ed. of Weisbach, Philada., 1848, article 107.)

21....... *Examples of the Application of Simpson's Rule to Earthworks.* In further illustration of this subject, suppose *Figs.* 72, 73, 74, and 75, to be cross-sections upon a railroad line, in stations of 100 feet, apart sections, with road-bed of 20, side-slopes 1 to 1, and other data *as dimensioned* upon the figures given; with equalizing lines properly drawn, reducing them to equivalent triangles, and with centre hights correctly ascertained.

Then, to find the End Areas to Intersection of Slopes.

$$
\begin{array}{llll}
& \text{Hights.} & \text{Widths.} & \text{Sq. Ft.} \\
Fig.\,72 = & 23\cdot4 & \times\ 47 & =\ 1100 \\
73 = & 27\cdot6 & \times\ 55\cdot5 & =\ 1532 \\
74 = & 28\cdot8 & \times\ 59\cdot9 & =\ 1725 \\
75 = & 27\cdot25 & \times\ 54\cdot6 & =\ 1488
\end{array}
\right\}
\begin{array}{c}
\text{Double Areas} \\
\text{in} \\
\text{Whole numbers.}
\end{array}
$$

Or, they may be computed, as is usual with engineers, by means of trapezoids and triangles, as they have been, indeed, in this case for the purpose of *verification*, and found to agree in whole numbers; there being, as usual, small differences in the decimal places.

When the ground surface is *irregular*, as shown in these cross-sections, the successive processes are *as follows:*

1. Find the equalizing line by *Art.* 8.

2. Ascertain the centre hight from intersection of slopes to equalizing line.

3. Find the extreme width, *or sum of distances out*, to the edges of tops of slopes, where they cut the equalizing line.

4. Find the *double areas* of the cross-sections, by multiplying together the hights and widths, or $h \times w$.

5. Find 8 *times the mid-section*, by means of *sum of* Hights $\times$ *sum of* Widths.

6. Then, for *Solidity*, proceed *prismoidally*, by Simpson's Formula as modified, for triangular solids.

The areas of the cross-sections having been duly verified, we may proceed to the calculation of some examples, *as follows:*

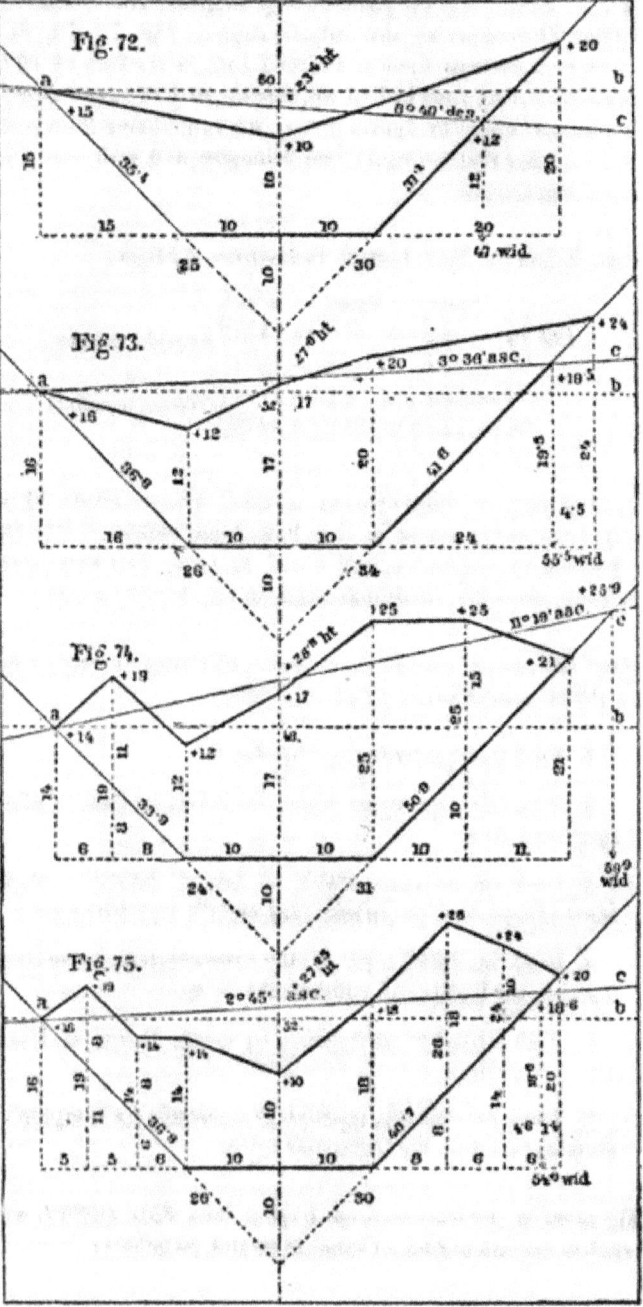

EXAMPLES.

Figs. 72 and 73.

Hights.   Widths.
23·4 × 47 = 1100 = Double Area of top.
27·6 × 55·5 = 1532 = " " base.
51 × 102·5 = 5228 = 8 times mid-section.
12)7860
655 = Prismoidal Mean Area.
100 Distance apart sections.
65500 = *Solidity* in Cubic Feet.

Figs. 73 and 74.

Hights.   Widths.
27·6 × 55·5 = 1532 = 2 *t.*
28·8 × 59·9 = 1725 = 2 *b.*
56·4 × 115·4 = 6509 = 8 *m.*
12)9766
814 = Prismoidal Mean.
100
81400 = *Solidity.*

Figs. 74 and 75.

Hights.   Widths.
28·8 × 59·9 = 1725 = 2 *t.*
27·25 × 54·6 = 1488 = 2 *b.*
56·05 × 114·5 = 6418 = 8 *m.*
12)9631
803 = Prismoidal Mean.
100
80300 = *Solidity.*

Cub. Ft.
Grade Prism to be deducted,
65500
Totalization 81400 to find the volume, from road-
80300 bed to ground.
227200 = *Sum of quantities.*

Grade Prism.
(Then, 227,200 — 30,000 = $\dfrac{197,200}{27}$ = 7304 Cubic Yards.

Tabulated by our 3d Method of Computation (Roots and Squares), the sum of the quantities, from *Fig.* 72 to *Fig.* 75 = 227,170 Cubic Feet (including Grade Prism); the slight difference of 30 Cubic Feet

arising from neglect of decimals on both sides ;—had these been carried further, the results would probably have been *identical*, or very
nearly so.

We may also *verify* this calculation by means of multipliers,
modelled after Simpson's, and applied to the areas, as given in the
examples, *as follows :*

Cross-sections figured in Nos. 72, 73, 74, and 75, stations 100 feet.

$$
\begin{array}{llcl}
& \text{Double} & & \\
\text{Sta.} & \text{Areas, etc.} \quad \text{Mults.} & & \text{Sq. Ft.} \\
72 & 1100 \times 0\text{·}5 = & & 550 \\
8 \text{ times mid-sec.} & 5228 \times 0\text{·}5 = & & 2615 \\
73 & 1532 \times 1 \;\; = & & 1532 \\
8 \text{ times mid-sec.} & 6509 \times 0\text{·}5 = & & 3255 \\
74 & 1725 \times 1 \;\; = & & 1725 \\
8 \text{ times mid-sec.} & 6418 \times 0\text{·}5 = & & 3209 \\
75 & 1488 \times 0\text{·}5 = & & \underline{744} \\
& & 6)& \overline{13630} \\
& & & \overline{2272}
\end{array}
$$

100 Double Interval.

Solidity, in Cubic Feet $= \overline{227,200}$, *same as before.*

The intervals are subdivided by the mid-sections into 50 feet
spaces, or *single interval.* The regular stations of 100 feet forming *a
double interval* in this case.

The Grade Prism being deducted (30,000 Cubic Feet), and the
remainder divided by 27, we have as before, *a volume of* 7304 *Cubic
Yards.*

22. *Observations upon Simpson's Rule.* SIMPSON appears to have
framed his rule for application to rectangular prismoids, and as such
he demonstrated it in reference to a diagram like *Fig.* 2, *Art.* **2**—
including of course those right triangles which are the halves of
rectangles.

He could have had no conception of the vast masses of earthwork
needed upon the public works of later days ; nor of providing a rule
for the mensuration of such ; nor, indeed, of the immense range the
Prismoidal Formula has since taken.

His rule (see *Art.* 2), though wonderfully flexible when applied to
rectangular or triangular figures, has no leading lines, common with

irregular ground; such surfaces then require to be *equalized*, by a single line on the principle of *Fig.* 14 *—converting the sections bounded by them into equivalent triangles before they can be computed by the Hights and Widths of Simpson's Rule, though we find occasionally that trapezium sections also, when not very much distorted, are often computable by the rule mentioned.

But, in applying such a rule to the rude masses of earthwork, so common at the present day, failing cases were to be expected, and the peculiar solid shown in *Figs.* 81 and 82 furnishes an example in point.

Figs. 81 and 82, Chap. V., *computed by Simpson's Rule.*

Hights. Widths.

$$
\begin{aligned}
60 \times 40 &= 2400 \\
30 \times 60 &= 1800 \\
\overline{90 \times 100} &= 9000 \\
&\ \ \ 12)\overline{13200}
\end{aligned}
$$

Prism. Mean Area = 1100
Common length . = 100

Solidity . . . . = 110,000 Cubic Feet.

But, by various examples, in *Arts* **29** and **30**, Chap. V., the *Solidity* = 130,000 Cubic Feet.

So that, in the case of this peculiar solid, *Figs.* 81 and 82, *Simpson's Rule* falls short = 20,000 Cubic Feet.

As the solid referred to has one end section *a Rhomboid*—the mid-section *a Pentagon*—and the other end *a Triangle.*

We could hardly expect Simpson's Rule, framed *for rectangular and triangular sections*, to answer in a case like this, and hence we mention it *especially.*

For all the solids which present sections, such as Simpson contemplated, his rule is *unquestionably correct*, while it is remarkably plain and simple in its application.

Further to illustrate what may be expected from Simpson's Rule, when applied by *equalizing lines* to rough and heavy sections, we will now compute the cases shown by *Figs.* 43 and 44, Chapter I.

Example, Illustrated by Fig. 43, Chapter I.

Side-slopes 1 to 1. No road-bed designated. *Proximate Computation*, by Simpson's Rule, to intersection of slopes; other dimensions as in *Fig.* 43.

Equalizing line of base $= b = 14°\ 2'$ asc.
"   "   top $= t = 15°\ 57'$ asc.

---

* *In substance*, this method is found in Hutton's Land Surveying (1770), quarto Mens.

Both these lines being drawn from the lowest side-hight, so as to *equalize* the areas, as per *Fig.* 14, Chapter I.

$$
\text{Areas}
\begin{cases}
1500 = b. \\
720 = t. \\
\text{Length, 100 feet.}
\end{cases}
$$

	Hights.		Widths.	
$b =$	37·5	×	80 =	3000
$t =$	25·7	×	56 =	1440
	63·2	×	136 =	8595·2

$$12)\ 13035·2$$

Prism Mean Area = 1086·3
Length . . . . = 100
Solidity . . . . = 108630
Same, by HUTTON = 108667
Difference . . . = − 37

Example, Illustrated by Fig. 44, *Chapter I.*

Side-slope 1½ to 1. No road-bed designated. *Proximate Computation,* by Simpson's Rule, to intersection of slopes, other dimensions as in *Fig.* 44.

$$
\text{Areas}
\begin{cases}
1352 = b. \\
726 = t. \\
\text{Length, 100 ft.}
\end{cases}
$$

Equalizing line of the base $b = 4° 30'$ asc.
　　"　　"　　"　　top $t = 1° 5'$ des.
　Both these lines being drawn from the lowest side-hight, so as to *equalize* the areas, as per *Fig.* 14, Chapter I.

Hights.		Widths.	
22·02	×	66 =	1453
29·81	×	90·7 =	2704
51·83	×	156·7 =	8122

$$12)\ 12279$$

Prismoidal Mean Area = 1023·25
Length . . . . . = 100
Solidity . . . . . = 102325
By Wedge and Pyramid = 102363
Difference . . . . . = − 38

With several other methods, this *proximate calculation* agrees within a few cubic yards.

Example from Warner's Earthwork, Art. 86.

A heavy embankment. For details, see Chapter V., near the close.

$$
\text{Areas}
\begin{cases}
2411 = b. \\
907 = t. \\
\text{Length, 100 feet.} \\
\text{Surface slope, } 15°.
\end{cases}
$$

$$\left\{\begin{array}{l}
\quad\quad\quad\text{Hignts.}\quad\text{Widths.} \\
\quad\quad\quad 36\cdot7 \times 131\cdot4 = 4822 \\
\quad\quad\quad 22\cdot5 \times 80\cdot6 = 1814 \\
\quad\quad\quad \overline{59\cdot2 \times 212\cdot0} = 12550 \\
\quad\quad\quad\quad\quad\quad\quad 12)\overline{19186} \\
\text{Prismoidal Mean Area . . } = 1599 \\
\text{Length . . . . . . } = 100 \\
\textit{Solidity} \text{ . . . . . . } = 159900 \text{ Cubic Feet.} \\
\text{For Cubic Yards} \div 27 \text{ . . } = 5922 \\
\text{Deduct vol. of Grade Prism } = 356 \\
\textit{Solidity} \text{ . . . . . . } = 5566 \text{ Cubic Yards.} \\
\text{By Hutton's Rule. . . . } = 5566 \\
\text{Difference . . . . . } = \pm 0
\end{array}\right\}$$

In calculating by *Simpson's Rule*, the example figured by *Figs.* 74 and 75—which agrees very nearly with HUTTON—we observe, by reference to the figures, that the ground slope at the end sections differs about 9°. So that we may safely assume that where the equalizing lines (representing the ground) have a nearly similar slope, and in the same direction, which do not differ more than 10° in their inclination, SIMPSON'S *Rule* may be safely used—this appears to be a *sure limit*, and we might perhaps go higher.

When the work happens to be upon uniform ground, or the equalizing lines have the same slope, as in the case cited from Warner's Earthwork, where the ground slope itself is uniform at 15°, the results obtained by *Simpson's Rule* ought to be exact, *and they appear to be so.*

CHAPTER IV.

THIRD METHOD OF COMPUTATION, BY MEANS OF ROOTS AND SQUARES; A PECULIAR MODIFICATION OF THE PRISMOIDAL FORMULA, WHICH WILL BE FOUND IN PRACTICE TO BE BOTH EXPEDITIOUS AND CORRECT, IN ORDINARY CASES.

23. . . . . . . This method of computation, by Roots and Squares,* appears to be the most rapid and compendious one treated by us, while it requires less data and preliminary work, and agrees in its results (for usual field work) with computations made direct by the Prismoidal Formula, of which, indeed, *it is only a special modification*, more concise and rapid in use, but at the same time *less accurate*. The formula for the Rule of Roots and Squares has been already described in the Preliminary Problems, *Art.* **10,** where it is numbered **XI.**, and is *as follows:*

$$\frac{h^2 + h'^2 + \left(\dfrac{h + h'}{2}\right)^2}{6} \times l = \mathrm{S}.$$

Where,

$h^2 =$ Representative square of area of top, from ground to intersection of slopes $= (t)$.

$h'^2 =$ Representative square of area of base, from ground to intersection of slopes $= (b)$.

$(h + h')^2 =$ Representative square of 4 times mid-sec. $= (4\,m)$.

$l =$ Distance apart sections—usually designated as (h) by the earlier writers, and hence continued by us to some extent; though l is clearly a more suitable symbol for earthwork, which, with a comparatively small cross-section, extends its length along the ground.

---

* This method is materially aided in its use by a good Table of Squares and Roots.— Prof. De Morgan's stereotyped edition of Barlow's Tables (8vo, London, 1860) is believed to be the *best*:—a very large edition was published, and this valuable work can be obtained from any of our importing booksellers *at quite a low price.*

When the numbers are large, the well known method of Logarithms gives the simplest process for *Involution or Evolution.*

108

Note.—That the hights of the end sections in this chapter are *always* to be considered as extending from the ground to intersection of slopes, or be representative of such.

The most important item in this notation is $(h + h')^2$, which, by geometry, we know to be equivalent to $4 \left(\dfrac{h + h'}{2}\right)^2$, while $\dfrac{h + h'}{2}$ is the representative in the mid-section of a line similar to h and h'.

So that this formula (for a single station) is, in fact, *equivalent to the Prismoidal Formula, as heretofore expressed, viz.:*

$$\frac{t + b + 4 m}{6} \times h = S,$$

but for *exact work* (our formula above) requires the end sections to be triangles, with a uniform ground slope.

Let us now apply the above formula to an entire cut or bank, to be computed by Hutton's Rule (adopted from Simpson)—see *Art.* **10,** Formula **IX.**

$$\text{Where } \frac{A + 4 B + 2 C}{6} \times Double\ interval = S.$$

Here, for a case of 6 *single* or 3 *double* intervals, as shown—*in the skeleton table*—below.

We have, for 3 double intervals or even spaces between stations *of equal length :*

$\Big\{$
$h^2 + h'^2$ . . . $= A$. The sum of extreme sections, each designating one end.

$3 (h + h')^2$ . . $= 4 B$. Mid-sections, standing on *even numbers.*

$2 (h')^2 + 2 (h)^2 = 2 C$. Regular Cross-sections, standing on *odd numbers.*

Double Interval = Any one of the uniform spaces, from 1 to 3, or 3 to 5, etc., being the *odd* numbers where the regular cross-sections stand.

$S =$ *Solidity* of entire cut of 3 *equal stations* in length.

Example 1. . . . . . Being a simple case (on irregular ground) of three uniform stations, or *double intervals,* of 100 feet each, the mid-sections falling in between, and dividing the length of 300 feet into *single intervals* of 50 feet each ; for which we will tabulate the example represented by *Figs.* 72, 73, 74, and 75, of Chapter III.—*in a skeleton table—as follows :*

STATEMENTS.	$\dfrac{h^2}{1}$	$(h+h')^2$	$\dfrac{h^2}{3}$	$(h+h')^2$	$\dfrac{h^2}{5}$	$(h+h')^2$	$\dfrac{h^2}{7}$
Regular stations designated by the numbers of the figures.	72		73		74		75
Places of mid-sections, on even numbers.		2		4		6	
Regular cross-section areas, upon the *odd* numbers.	550·		766·		862·5		744·
Square roots of areas of regular cross-sections.	23·45		27·68		29·37		27·28
Sums of square roots.		51·13		57·05		56 65	
* Squares of sums, or 4 times the proper mid-section.		2615·		3255·		3269·	
		Extra decimals thrown together here.					

Having given the skeleton table of *data*, we will now tabulate for *solidity* on three different plans, any one of which may be adopted, or in fact any other which truly represents the formula given.

Tabulation for Solidity.

On the plan of *Art.* **10**, in Chapter I.	By Simpson's Rule (as given by Hutton).	By Multipliers, modelled after Simpson's.
Sta. 72 Areas . . = 550	A. 4 B. 2 C.	End areas, and 4 times mid-section. Mults. Results
4 mid-sec. . . 2615	550 2615 766	550 × 1 = 550
73 . . . . ·{ 766 / 766	744 3255 706	2615 × 1 = 2615
4 mid-sec. . . 3255	——	766 × 2 = 1532
	1294 3209 862·5	3255 × 1 = 3255
74 . . . . { 862·5 / 862·5	4 B = 9079 862·5	862·5 × 2 = 1725
4 mid-sec. . . 3209	2 C = 3257 3257	3209 × 1 = 3209
75 744	A = 1294	744 × 1 = 744
6)13630	6)13630	6)13630
General Mean Area = 2271·7	2271·7	2271·7
Double Interval. . = 100	100 Double Int.	Double Interval . . = 100
Solidity in C. Feet = 227,170	*Solidity* = 227,170 in C. Feet.	*Solidity* in C. Feet = 227,170
Whole length of cut 300 feet.	Whole length of cut 300 feet.	Whole length of cut 300 feet.

24. Now, for further illustration :— Take any cut or bank—say of 6 (or any *even* number of) *equal* stations—their termini being tem-

---

* HUTTON and other geometers have shown that the square of any line equals 4 times that of half the line;—and that similar triangles are to each other *not only* as the squares of their like sides, *but also as the squares* of any *similar lines ;* and these principles of Geometry lay at the foundation of the method of computation, developed in this Chapter IV. (as already indicated in the Preliminary Problems).

porarily numbered in the series of *odd* numbers, while the intermediate spaces (or places of mid-sections) are also temporarily numbered
in the series of *even* numbers, and the places of cross-sections and mid-
sections, as well as those of the symbols used in the formula, all
regularly marked, *as follows:*

Regular stations.	1		3		5		7		9		11		13
Places of cross-secs.	⊙		⊙		⊙		⊙		⊙		⊙		⊙
" mid-secs.		2		4		6		8		10		12	
Symbols of formula.	h^2	$(h+h')^2$	h^2	$(h+h')^2$	h^2	$(h+h')^2$	h^2	$(h+h')^2$	h^2	$(h+h')^2$	h^2	$(h+h')^2$	h^2

This little skeleton table shows the positions of the representative
squares equivalent to the areas of the several regular cross-sections
computed, and also of 4 times the proper mid-sections, which belong
between them, and it will indicate the manner in which they are
combined relatively to the odd numbers, which represent the regular
stations; so that having computed the regular cross-sections, we can
readily assemble them in a skeleton table, compute from them by
Roots and Squares the other data demanded by the formula, and
proceed to tabulate for *Solidity,* as has been already shown, and will
be more conspicuously exhibited hereafter.

Upon the foregoing principles we will now proceed with an entire
piece of heavy embankment, succeeded by a rock cut, as shown in
the annexed, *Fig.* 76.

Example 2. . . . BANK = 1000 feet long. . . . *Fig.* 76.

Skeleton Table of Data, Given or Computed.

Length of regular stations 100 feet—intervals produced by Mid-sections 50 feet.

Regular stations of 100 feet =	1	2	3	4	5	6	7	8	9	10	11
Temporary numbers . . . =	1	3	5	7	9	11	13	15	17	19	21
Regular Cross-section Areas =	24	185	495	1467	3123	3123	3123	1978	1197	391	24
Places of mid-secs., Inter- mediates at 50 ft. (really), } =		2	4	6	8	10	12	14	16	18	20	
½ Roots of the Cross-sec- tion Areas . . . . . } =	4·90	13·60	22·25	38·30	55·88	55·88	55·88	44·47	34·60	19·77	4·90
Sums of Roots . . . . =		18·50	35·85	60·55	94·18	111·76	111·76	100·35	79·07	54·37	24·67	
Squares of Sums, or 4 times } the Mid-section Areas. } =		342·25	1285·72	3666·30	8869·87	12490·30	12490·30	10070·12	6252·06	2956·10	608·61	

* For *Figs.* 77 and 78, illustrating a supposed basis of the Prismoidal **Formula,** and
its connexion with Simpson's Rule for Cubature (see Chap. VII.).

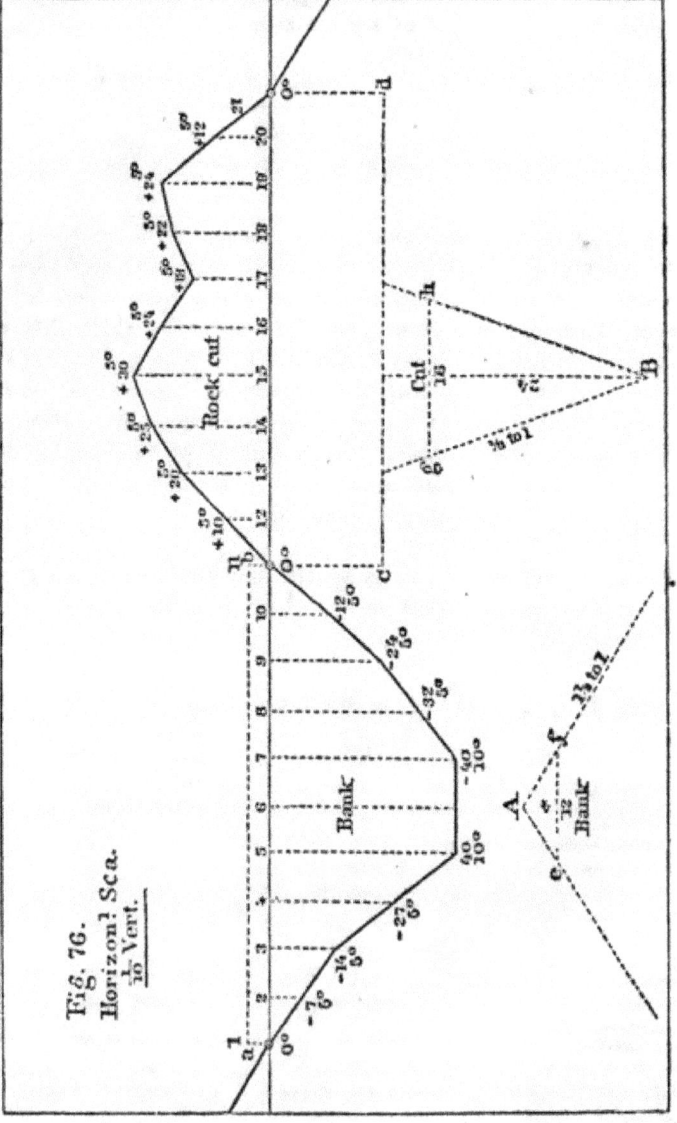

Fig. 76.
Horizonᵗˡ Sca.
$\frac{1}{10}$ Vert.

Tabulations for Solidity ;

By 100 feet stations, or 50 feet intervals.

1.

Regular stations of 100 feet.	Cross-section Areas.
1 . . . . . =	24
4 times mid-section . . =	342·25
2 . . . . . = {	185
	185
" " . =	1285·82
3 . . . . . = {	495
	495
" " . =	3666·30
4 . . . . . = {	1467
	1467
" " . =	8869·87
5 . . . . . = {	3123
	3123
" " . =	12490
6 . . . . . = {	3123
	3123
" " . =	12490
7 . . . . . = {	3123
	3123
" " . =	10070·12
8 . . . . . = {	1978
	1978
" " . =	6252·06
9 . . . . . = {	1197
	1197
" " . =	2956·10
10 . . . . . = {	391
	391
" " . =	608·61
11 . . . . . =	24
	6)89243·13

Gen.mean area to int.of slopes = 14874

100

Solidity in c. ft.to int.of slopes = 1487400 of
BANK.

2. By Multipliers, modelled after Simpson's.

Mults.	Results.
1 . . . . =	24
1 . . . . =	342
2 . . . . =	370
1 . . . . =	1285
2 . . . . =	990
1 . . . . =	3667
2 . . . . =	2934
1 . . . . =	8870
2 . . . . =	6246
1 . . . . =	12490
2 . . . . =	6246
1 . . . . =	12490
2 . . . . =	6246
1 . . . . =	10070
2 . . . . =	3956
1 . . . . =	6252
2 . . . . =	2394
1 . . . . =	2956
2 . . . . =	782
1 . . . . =	609
1 . . . . =	24

Proof :   6)89243
Gen.mean area to int.of slopes = 14874

100

Solidity in c.ft. to int.of slopes = 1487400 of
BANK.

Example 2—Continued.   ROCK CUT = 1000 feet long. . . *Fig. 76.*

Skeleton Table of Data, Given or Computed.

Length of regular stations 100 feet ; which, by means of the Hypothetical Mid sections, cover the ground with 50 feet intervals.

Regular stations of 100 feet =	11	12	13	14	15	16	17	18	19	20	21
Temporary numbers . . . =	1	3	5	7	9	11	13	15	17	19	21
Regular Cross-section Areas =	192	386	646	801	975	768	549	706	771	453	192
Places of mid-secs., intermediates at 50 ft. (really). } =	2	4	6	8	10	12	14	16	18	20	
½ Roots of the Cross-section Areas . . . . } =	13·86	19·65	25·42	28·31	31·23	27·71	24·27	26·57	27·77	20·81	13·86
Sums of Roots . . . . . =	33·51	45·07	53·73	59·54	58·96	51·98	50·84	54·34	48·58	34·67	
Squares of Sums, or 4 times the Mid-section Areas. } =	1122·97	2031·30	2886·91	3545·01	3476·28	2701·92	2584·70	2952·83	2360·01	1202·01	

8

Tabulations for Solidity:

By 100 feet stations, or 50 feet intervals.

1.

Regular stations of 100 feet.	Cross-section Areas.
11 . . . . . =	192
4 times mid-section . . . =	1122·92
12 . . . . . = {	386 / 386
" " . =	2031·30
13 . . . . . = {	646 / 646
" " . =	2886·91
14 . . . . . = {	801 / 801
" " . =	3545·01
15 . . . . . = {	975 / 975
" " . =	3476·28
16 . . . . . = {	768 / 768
" " . =	2701·92
17 . . . . . = {	589 / 589
" " . =	2584·70
18 . . . . . = {	706 / 706
" " . =	2952·83
19 . . . . . = {	771 / 771
" " . =	2360·01
20 . . . . . = {	433 / 433
" " . =	1202·01
21 . . . . . =	192
	6)37397·89

Gen.mean area to int.of slopes = 6233
       100

Solidity in c.ft.to int. of slopes = 623300 of ROCK CUT.

2. By Multipliers, modelled after Simpson's.

Mults.	Results.
1 . . . . =	192
1 . . . . =	1123
2 . . . . =	772
1 . . . . =	2031
2 . . . . =	1292
1 . . . . =	2887
2 . . . . =	1602
1 . . . . =	3545
2 . . . . =	1950
1 . . . . =	3476
2 . . . . =	1536
1 . . . . =	2702
2 . . . . =	1178
1 . . . . =	2585
2 . . . . =	1412
1 . . . . =	2953
2 . . . . =	1542
1 . . . . =	2360
2 . . . . =	866
1 . . . . =	1202
1 . . . . =	192

Proof :   6)37393

Gen.mean area to int.of slopes = 6233
       100

Solidity in c.ft. to int.of slopes = 623300 of ROCK CUT.

25. In the preceding example, the side-slopes of the BANK are 1½ to 1 — road-bed = 12; while in the ROCK CUT, the side-slopes are ¾ to 1 — road-bed = 16; and in all these calculations (we repeat), *the sectional areas*, in every case, are taken from ground line to intersection of side-slopes; and *the hights*, from the vertex of the common angle thus formed to the line, or lines, representing the surface of the ground.

So that in all such computations—if the contents above or below a given road-bed be desired in the results, then the volume of the grade prism (being included in the summation) must in every case *be duly deducted*.

The volume of the grade prism depends upon its sectional area, and the length of the bank or cut—these calculations are very simple, and once made, remain unchanged as long as the road-bed and side-slopes *continue uniform*.

Geometers having shown that the areas of similar triangles are to each other, not only as the squares of like sides, but also as the squares of *any similar lines* in each, and these often occurring in earthwork solids, when their cross-sections are converted into triangular areas, by the prolongation (to a junction) of the side-slopes, it becomes of importance *to classify* the relations existing among lines and their squares, as well as the squares and rectangles of their sums and differences;—this has been well done in J. R. Young's Geometry (London, 1827), in several successive propositions:—Book II., 4, 5, 6, 7, and 8.

Now, suppose any line to be divided into *two parts*, h and h'—then, by these propositions, *we have:*

1. $(h + h')^2 = 2(h + h') \times \left(\dfrac{h + h'}{2}\right).$
2. $(h + h')^2 = h^2 + h'^2 + 2hh'.$
3. $(h - h')^2 = h^2 + h'^2 - 2hh'.$
4. $h^2 - h'^2 = (h + h') \times (h - h').$
5. $h^2 + h'^2 = \frac{1}{2}(h + h')^2 + \frac{1}{2}(h - h')^2.$
6. $2(h^2 + h'^2) = (h + h')^2 + (h - h')^2.$

As these lines, or parts of lines, may, and often do, occupy in similar triangles the relation of *like lines*, they become of some consequence in earthwork calculations, and in various forms can be traced through many of the formulas now before the public.

We will now give an example from Warner's Earthwork (*Art.* 124), to show that small variances may be expected in employing the Rule of this Chapter upon irregular ground:—indeed, it is only in uniform sections that an exact agreement of Rules can be anticipated, but the variations (always small) are not unlikely to balance themselves in computing considerable lengths of line.

$$
\text{Here,}
\begin{cases}
\begin{cases}
\text{End areas to grade} \dots \dots = 846\cdot5 \dots = 915.5 \\
\text{Grade Triangle to add} \dots \dots = 196 \dots = 196 \\
\text{End areas to int. of slopes} \dots = \overline{1042\cdot5} \dots = \overline{1111\cdot5}
\end{cases} \\
\text{Square Roots} \dots \dots \dots = 32\cdot29 \dots = 33\cdot34 \\
\text{Sums of Roots} \dots \dots \dots = 65\cdot63 \\
\begin{cases}
\text{Square of sum, or} \\
\text{quadruple mid-section} \dots \dots = 4308
\end{cases}
\end{cases}
$$

Length, 100 feet.

Then, Prismoidally,

$$
\begin{array}{ll}
\text{Sum end areas} \dots \dots \dots & = 2154 \\
\text{Quadruple Mid-section} \dots \dots & = 4308 \\
& \overline{6)6462} \\
& \overline{1077} \\
\text{Length} \dots \dots \dots \dots & = \quad 100 \\
& \overline{107700} \\
\text{Off Grade Prism} \dots \dots \dots & = \quad 19600 \\
& \overline{27)88100} \\
\textit{Solidity} \text{ in Cubic Yards} \dots \dots & = \quad 3263
\end{array}
$$

As computed by Warner (3274, C. Y.); and also by Hutton's General Rule (3274, C. Y.), the difference made by our Rule of this Chapter is, 11 Cubic Yards, *or about ⅓ of one per cent.*

Comparison of the method of this Chapter with the test examples of Chapter II., as computed by Hutton's General Rule (each for 100 feet in length).

1. *Three-level Ground.*

(See *Figs.* 53, 54, and 55.)   C. Yards.

Computed by Roots and Squares (method of this Chapter) = 2337·6

" " Hutton's General Rule (Chapter II.) . . . = 2339.6

Difference . . . . = — 2

2. *Five-level Ground.*

(See *Figs.* 56, 57, and 58.)   C. Yards.

Computed by Roots and Squares (this Chapter) . . . . = 1061·1

" " Hutton's General Rule (Chapter II.) . . . = 1061·1

Difference. . . . . = 0

3. *Seven-level Ground.*

C. Yards

Computed by Roots and Squares (this Chapter) . . . . $= 1990\cdot$
"   "   Hutton's General Rule (Chapter II.) . . . . $= 1989\cdot6$
$$\text{Difference} . . . . = + 0\cdot4$$

4. *Nine-level Ground.*

C. Yards.

Computed by Roots and Squares (this Chapter) . . . . $= 2562\cdot9$
"   "   Hutton's General Rule (Chapter II.) . . . . $= 2562\cdot9$
$$\text{Difference} . . . . = 0$$

We will now give another example from Warner's Earthwork, computed by the method of this chapter.

Heavy Embankment (Art. 86).

Areas . . . . . $= 2411$ 907
$\sqrt{\text{Roots}}$ . . . . $= 49\cdot10$ $30\cdot12$
Sums of Roots . . . . . $= 79\cdot22$
Square of sum,
  or quadruple $\Big\}$ . . . . $= 6276$
  mid-section.

Then, Prismoidally,

$\begin{cases} \text{Sum of ends . . . } = 3318 \\ \text{Quadruple Mid-sec. } = 6276 \\ \qquad\qquad\quad 6)\overline{9594} \\ \times \text{ length . . . . } = \overline{159900} \\ \div 27 \text{ for C. Yards } = \quad 5566 = \text{Same as Hutton's Gen. Rule.} \end{cases}$

From the above it will be observed that, with a Table of Powers and Roots at hand, *the method of this chapter affords a very convenient and speedy test for volumes, found by other processes, and it is a proximately correct one.*

CHAPTER V.

26. . . . . . . Sir John Macneill (1833) hath shown that a Prismoid of Earthwork is really a prism with a wedge superposed (as we have already mentioned in *Art.* **4**)—that the wedge is also divisible into two pyramids—and that the formulas for volume, in these three chief bodies of solid geometry, form, by addition, *the Prismoidal Formula.*

Regarding the Prismoid in this way, and assuming it to have been diagrammed as shown in *Fig.* 8, *Art.* **6** (both end sections upon one drawing), it is easily computable *when reduced to a level on the top,* and the back of the wedge is a trapezoid, by means of Formula **VI.,** *Art.* **6.**

This Formula is:

$$\frac{(B + b + b) \times (H - h)}{6} + (h^2 r - \text{Grade Triangle}) \times l = Solidity,$$

to road-bed, and omitting G. T. to intersection of slopes.

Where,

B   = Top-width of back, or larger parallel side of trapezoid, measured horizontally.

b   = Bottom-width of back, or lesser side of trapezoid, equal also to the edge, which is the horizontal top-width of smaller end section, at a distance forward = to the common length of wedge and prism.

H and h = Vertical hights of the end sections to intersection of slopes.

H — h = Hight of back of wedge.

r = Ratio of side-slopes to unity, or cot. of slope angle.

$h^2 r$ = Area of prism to intersection of slopes, and less Grade Triangle = area of section from ground to road-bed.

118

In calculating by this Formula we may omit the Grade Triangle if we choose (though we should have to supply a more complicated expression for $h' r$), and might, perhaps, somewhat simplify the computation thereby; but *if* used in *area*, we must be careful to account for it in *volume;* while the hights need only be extended from ground to road-bed; though as *their difference only* is used here, that is not material—*and altogether we would gain so little by the change as to make it unadvisable.*

In words, this Formula may be expressed *as fol-lows:* $\Big\}$ *(Mean Area Wedge + Mean Area of Prism) × Common Length = Solidity, of the Prismoid, to intersection of slopes, and minus G. T. to Road-bed.*

Inasmuch, however, *as a trapezoid is always reducible to an equivalent rectangle,* we may consider this matter of the superposed wedge in a more general manner, without the necessity of first reducing the trapezoidal, or triangular, cross-section to a level on the top, or slope of $0°$.

Before entering upon this branch of the subject we may, however, state that the reason why, in a wedge with a trapezoidal back, we sum up all the three parallel sides of back and edge × by hight of back ÷ by 6, and finally multiply by length *for volume*—is drawn from the common rule for a wedge—(Twice width of back + edge × by hight of back ÷ by 6, and × by length = *Volume.*) But in a wedge with a *trapezoidal* back—the $\frac{1}{2}$ sum of top and bottom parallel sides × 2 = simply *the sum of those parallel sides;* and, as in an earthwork solid, the lesser parallel side also (*generally*) equals the edge, that being the top line of the smaller end section, situated at a distance of the length *forward.* Hence, B + b + b is usually equivalent to $\frac{B + b}{2} × 2 + (b$ the length of the edge)—which will be found in substance as a term in Hutton's Rule for wedges (4to Mens., 1770); but more concisely expressed in Chauvenet's Theorem.

References to Fig. 79.*

a d = End view of the back of a rectangular wedge.

a f = Equivalent parallelogram, of which *a g* is the base. and *a* D *the altitude.*

---

a D   = Horizontal projection (70·71), or width of $a b$ (the back).

$a l$ = Horizontal projection (35·36), or width of $a h$ (the edge)

$a e g k$ = The initial square of 50 square feet area, which is con-

tained in the back = $\dfrac{707}{50}$ = 14·14 times.

A B $\{$ Vertical and horizontal
C D $\{$ rectangular axes.

Fig. 79.

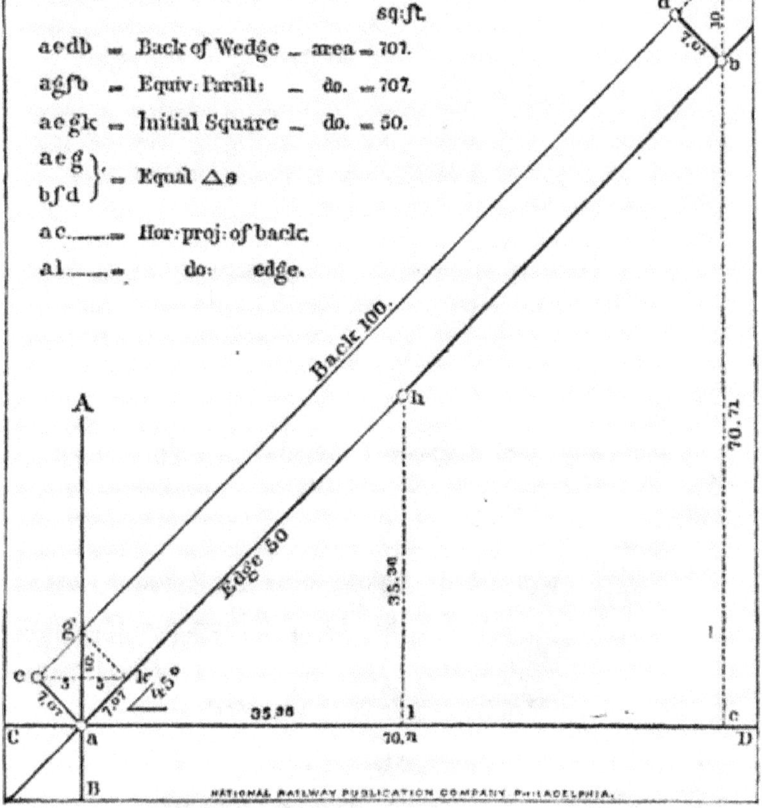

sq:ft.

acdb – Back of Wedge – area – 707.

agfb – Equiv: Parall: – do. – 707.

aegk – Initial Square – do. – 50.

$\left.\begin{matrix} a e g \\ b f d \end{matrix}\right\}$ – Equal △s

ac......– Hor:proj: of back.

al......–   do: edge.

The triangles, $a e g$ and $b d f$, are *identical*, and the one cut off, and the other added, make the two parallelograms, $a d$ and $a f$, *precisely equivalent* = 707 area, for each.

$a\,b =$ Width of back of rectangular wedge, inclined at an angle
of 45° $= 100$.

$a\,h =$ Width of edge, or top of forward, or smaller, section $= 50$.

Now (as above mentioned), *a trapezoid being always reducible to an equivalent rectangle*, we may consider in this place the superposed wedge (with reference to *Fig.* 79), without the necessity of first equalizing the end cross-sections, by level lines on the top, as will be more clearly seen further on.

However much the back or edge of a rectangular wedge may be inclined from a level plane, the resulting volume is still the same by using their projections upon the horizontal one of two rectangular axes (as C D), instead of the actual widths of back or edge, whilst the hight of the back becomes the base of an equivalent parallelogram, of which the projection is the altitude ;—this will become evident by reference to *Fig.* 79.

For example, let us now compute the wedge shown in the figure: 1st, As though it were upon a level, and the back a rectangle.   2d, As an oblique parallelogram on the back, and inclined at 45° from a level line.

1. *Rectangular back*—supposed to be level.   Length of wedge $=$ 100.   Breadth of back $= 100$.   Edge $= 50$.   Hight of back $=$ 7·071.

Here we have :—Sum of the 3 parallel sides of edge and back $\div$ 3.

$$\left\{\begin{array}{l} \left.\begin{array}{r} 100 \\ 100 \end{array}\right\} = \text{Back.} \\ 50\ \ = \text{Edge.} \\ \overline{3\,)250} \\ 83\tfrac{1}{3} = \text{Average multiplier} \end{array}\right. \qquad \textit{Right Section} \left\{\begin{array}{l} 7\cdot071\ \ = \text{Altitude.} \\ 100 = \text{Length.} \\ \overline{2\,)707\cdot100} \\ 353\cdot55 \\ 83\tfrac{1}{3} \end{array}\right.$$

. . . $=$

$$\textit{Volume} \ = \overline{29,463} = \text{C. Feet.}$$

Computed after Chauvenet's Theorem (Geom., VII. 22).

2. *Oblique-angled Parallelogram for Back,* and inclined 45°. Length of wedge.= 100.   Hight of back = 10.   Horizontal projection of back = 70·71.   Horizontal projection of edge = 35·36.

$$\frac{\text{Sum of the 3 parallel sides or edges}}{3} =$$

$$
\left\{
\begin{array}{l}
70\cdot71 \\
70\cdot71 \\
35\cdot36 \\
\overline{3)176\cdot78} \\
\overline{58\cdot927}
\end{array}
\right.
\quad
\begin{array}{l}
\left.\begin{array}{l}\\ \end{array}\right\} = \text{Back.} \\
= \text{Edge.} \\
 \\
= \text{Average multiplier .}
\end{array}
\quad \textit{Right Section}
\left\{
\begin{array}{l}
10 \qquad= \text{Altitude.} \\
100 \qquad= \text{Length.} \\
\overline{2)1000} \\
\overline{500} \\
= 58\cdot927
\end{array}
\right.
$$

$$Volume = \overline{29{,}463}$$

It is evident, from a consideration of the above case of a rectangular wedge, whether level or inclined, that the same process would apply to the trapezoidal wedge (usual in earthworks), either by its reduction to an equivalent rectangular one, or (when diagrammed together) by projecting both sides of the back, and also the edge, upon the horizontal axis, and ascertaining the respective lengths of these three projections, to be used in the computation of volume, by Chauvenet's Theorem,[*] *instead of their actual measured lengths,*—this is in fact the method of the engineer, who usually disregards the inclination of the ground, and takes all his measures horizontally and vertically.

The *hight* of the back of the inclined wedge being in the case above, ascertained by dividing the known area of the back of the rectangular wedge, by the Arithmetical Mean of the horizontal projections of its top and bottom breadths;—both *equal* in the above rectangular back, but always *unequal* in a trapezoidal one.

With these preliminary observations, we will now give the rule for finding the volume of the superposed wedge in ordinary earthworks, with examples to show how, by the simple addition of the under-prism, the solidity of the entire earthwork, between any two cross-sections of given area, and distance apart, *is easily ascertained,* in all cases, within a *limit* hereafter discussed (*Art.* **29**).

27. . . . . . . *Rules for Computation by Wedge and Prism.*   The data required to be given will be *as follows:*

---

[*] Chauvenet's Geom., VII. 22 (Philada., 1871).

1. Areas of end cross-sections.

2. Distance apart, or common length of wedge and prism.

3. Sum of distances out, to ground edges of side-slopes,—which are, in fact, the projections or horizontal widths of back and edge, as well as the right and left distances of the field engineer.

The first is obtained by well-known processes, and the two latter are always supplied by the Field Book of the engineer.

Then, as preliminary steps: (1) Find the difference of the areas of the end cross-sections, *which difference* is the area of the back of the superposed wedge. (2) Divide this difference of area by half the sum of the widths of the back (or horizontal projections), which gives the vertical mean hight of the back. Now, the lower side of the back (when both sections are diagrammed together) equals the edge (or top-width of the smaller end section) supposed to be *forward*, at a distance equal to *the common length*. So that if B = top-width of larger end section, — b will equal its bottom width (*and also that of the edge*)—so that $B + b + b$, for the wedge-shaped part, would give the sum of the three parallel edges (or, in reality, their horizontal projections) to be divided by 3, *for use in Chauvenet's Theorem.*

RULE.—When the width of the large end *is equal to or greater than* that of the small one.

1.

$$\frac{\text{Vertical mean hight} \times \text{distance apart sections}}{2} \times$$

$$\frac{\text{Sum of the three parallel edges}}{3} = \textit{Volume of Superposed Wedge.}$$

2. Smaller end area $\times$ length (or distance apart sections) = *Volume of Prism.*

These two results, added together = *Solidity of the whole Prismoid.*

a. . . . . . . Prior to giving examples in illustration of our rule, it appears necessary in this place to make some explanations to show the generality of the application of the rule drawn from Chauvenet's Theorem (Geom., VII. 22) *for the volume of wedges.*

Wedges are always formed by the truncation of triangular prisms, which may be termed *their elementary body;* and are usually designated by the outlines of their backs—as Rectangular, Triangular, Trapezoidal, etc.—*The Initial Wedge* may be assumed to have *a square back;* by successive transformations of which, several varieties are easily formed.

(1) Let the back of a rectangular wedge (*or the initial wedge*) be a square, on a side of 6, edge 12, length 20.—Then, the right section $= (6 \times 20) \div 2 = 60.$—One-third of the sum of the lateral edges $= (6 + 6 + 12) \div 3 = 8;$ and $60 \times 8 = 480 =$ *Volume of the Square Wedge.*

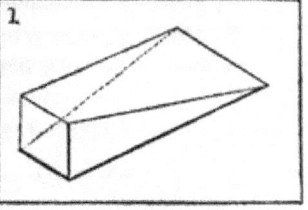

(2) Now, suppose the edge of (1) to be contracted to *a point;* then, the wedge becomes a pyramid, for which case the rule *also holds;*—thus, right section $=$

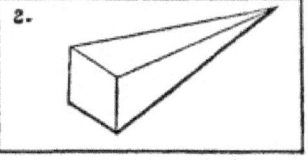

$60 \ldots \ldots \frac{1}{3}$ sum of edges $= (6 + 6 + 0) \div 3 = 4;$ and $60 \times 4 = 240 =$ *Volume.*

Proof: By the common rule for pyramids, we have, base $(6 \times 6) \div 3 = 12;$ and $\times$ by altitude $20 = 240 =$ *Volume,* the same as before.

(3) Suppose the back of the square wedge (1) to be converted into an isosceles triangle, on a base of 6, and hight of 6— other dimensions as in (1)— then right section $= 60 \ldots \ldots$ $\frac{1}{3}$ sum of edges $= (6 + 0 + 12) \div 3 = 6;$ and $60 \times 6 = 360 =$ *Volume.*

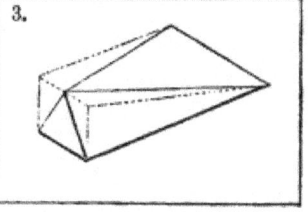

Proof: Now, the inscription of the isosceles triangle, *within the square back*, evidently cuts off two pyramids, of which the *volume* of each $= (3 \times 6) \div 2 = 9 \div 3 \times 20$ length $\times 2$ in number $= 120$ *Volume*, of pyramids cut away from the square wedge (1); —then, $480 - 120 = 360 =$ *Volume,* the same as before.

(4) Now, suppose (1) and (2) to be placed in contact *sidewise*, then they form together a rectangular wedge, back, 12 by 6; edge, 12; length, 20 :—right section $= 60 \ldots \ldots \frac{1}{3}$ sum of edges $= (12 + 12 + 12) \div 3 = 12;$ and $60 \times 12 = 720 =$ *Volume.*

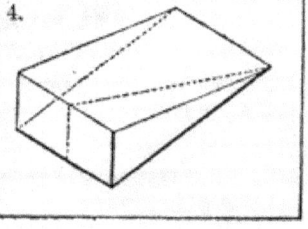

Proof: *By two Pyramids* = (72 ÷ 3 × 20 = 480) + (60 ÷ 3 × 12 = 240) = 720, *the same Volume;* or, by addition of (1) and (2) = 480 + 240 = 720, *Volume* as before.

(5) Suppose now the vertical sides of the square back of (1) to close in gradually until they meet and coincide in a single vertical line; then the back has vanished, and become a vertical edge, while the original one remains horizontal, *dimensioned*

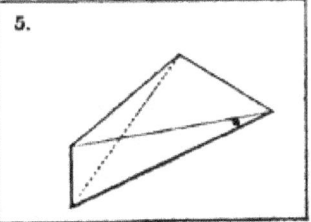

along with the other parts as in (1)—and we have right-section 60 . . . . . . ⅓ sum of edges = (12 + 0 + 0) ÷ 3 = 4; and 60 × 4 = 240 = *Volume* of this peculiar double-edged wedge; which is composed *of,* or may be decomposed *into,* two pyramids, based on the right-section, as common to both, and each having an altitude of half the edge, or 6 (though such equal division of edge is not essential); hence, we may assume the edge 12 to be a double altitude; and $\left(\dfrac{60}{3} \times 12\right)$ = 240 = *Volume of both—* the same as before.

(6) Now, suppose the vertical sides of the square (1) to become inclined (at any angle that will not extinguish the base of the back), say at an angle of ⅓ to 1 side-slope, thus reducing the base from 6 to 2, then we have the right-section as before = 60 . . . . . . ⅓

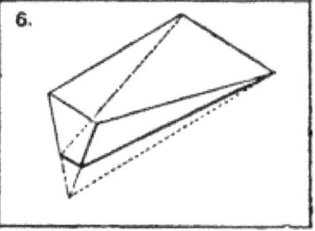

sum of edges = (6 + 2 + 12) ÷ 3 = 6⅔; and 60 × 6⅔ = 400 = *Volume of Trapezoidal Wedge.*

Proof: In this case two triangular pyramids are cut away from the original solid, by the sloping sides, having *together* a base of 4, and altitude of 6; then, (6 × 4) ÷ 2 = 12, which ÷ 3 and × 20 common length = 80 Volume cut away—but Volume of (1) = 480 — 80 = *residual Volume* = 400, as before.

(7) Now, suppose two sides of the square back of (1) to gradually reduce their contained angle, and finally to vanish upon the

diagonal—then the back becomes a right-angled triangle
(the side joining the right-angle,
say perpendicular to the edge),
and this wedge has *two edges* (one
original, and the other now
formed at the side connecting

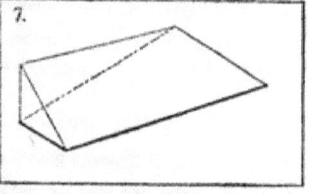

with the acute angle, both being horizontal edges). Then, the
right-section = 60 . . . . . . ⅓ sum of edges (6 + 0 + 12) ÷ 3
= 6; and 60 × 6 = 360 = *Volume.*

Proof: Divided by a plane *diagonally* through the vertex of
the triangular back, and opposite corner of the edge, we may
decompose this wedge into two pyramids—the one with a base
= the right-section = 60, and altitude = the original edge =
12; then, 60 × 12 ÷ 3 = *Volume* . . . . = 240

The other, with a base equal to the triangular back, or
(6 × 6) ÷ 2 = 18, and an altitude = the length = 20;
then, 18 ÷ 3 = 6, and × length 20 = *Volume* . . . = 120

Total Volume of both Pyramids . . . . = 360
the same as before.

(8) *A Rhomboid Wedge* is
computed in a similar manner:
—thus, let the rhomboidal back
have a vertical diagonal = 12;
the other = 4; an edge of 12;
length = 20; and the side-slopes
being ⅓ to 1.

Then, the right-section =
$\dfrac{12 \times 20}{2}$ = 120 . . . . . . ⅓ sum of edges, $\dfrac{4 + 12 + 0}{3}$ = 5⅓; and

120 × 5⅓ = 640 = *Volume.*

Now, by cutting off from the rhomboid, near the lower angle,
any given triangle, we have remaining *a Pentagonal Wedge.*

Thus, suppose we cut off a triangular wedge having the base
of its back uppermost = 2; altitude = 3; common length and
edge = 20 and 12.

Then its right-section = $\dfrac{3 \times 20}{2} \times \dfrac{2 + 12 + 0}{3}$ = 140 *Volume, cut off.* And 640 — 140 = 500 = *the Volume of the residual
Pentagonal Wedge.*

(9) Let us now consider a *Trapezoidal Wedge*—dimensioned like (8), with side-slopes of $\frac{1}{3}$ to 1, forming the top of the back, while its base = 2.

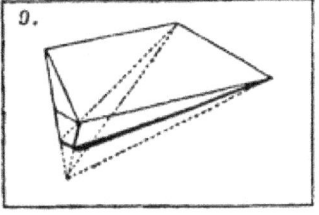

Let one side-hight = 12 above intersection of slopes; the other = 6; the edge = 12; and the length = 20.

Now, we may compute this wedge in two parts *as follows:*

1. As a triangular wedge, above the level of the lowest side-hight.

$$\left(\frac{6 \times 20}{2}\right) \times \frac{4 + 12 + 0}{3} \quad \ldots \ldots = 320$$

2. As a trapezoidal wedge, between the level mentioned and the base of the back.

$$\left(\frac{3 \times 20}{2}\right) \times \frac{4 + 2 + 12}{3} \quad \ldots \ldots = \underline{180}$$

Total Volume . . . . = 500

Or, as in (8), we may compute the body as a *Rhomboidal Wedge,* and deduct the triangular wedge cut away below the base of 2,—as in fact we did in (8),—*the resulting volume* being 500, the same as herein found.

Finally, we perceive *that from* (1) *the square or initial wedge* we may easily deduce several varieties of wedges, and might go further.

After this necessary digression, indicative of the simplicity, generality, and value of Chauvenet's Theorem, we will now proceed to illustrate our own rule (deduced from this theorem), as applied to Earthworks, by several examples.

28. . . . . . . Here follows the calculation of some examples.

Example 1.—Computation by Wedge and Prism, tested by Hights and Widths, under Simpson's Rule

References to Fig. 80.

In this case equal slopes of 1 in 4 form *a ridge* in the larger end section, and *a hollow* in the lesser one.

Dimensioned as shown in the figure annexed.

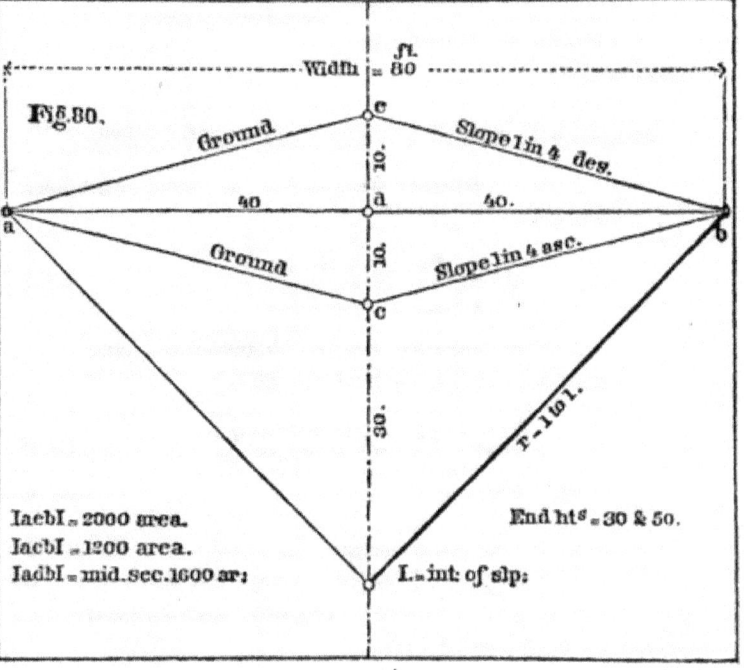

Data.

Sq. Ft.
$$\left\{\begin{array}{l}\text{Differences of areas of end sections}\dots\dots\dots\ = 800\\ \text{Widths, or horizontal projections, equal for both sections}\ \ .\ = 80\\ \text{Distance apart sections}\ \dots\dots\dots\dots\dots\ = 100\end{array}\right.$$

To find the vertical mean hight of back of wedge.

$$\text{End Areas} = \left\{\begin{array}{c}2000\\1200\end{array}\right\}\ \textit{Difference of Areas.}$$

Half sum of widths $= 80)\ \overline{800}$

$\qquad\qquad\qquad 10 =$ Vertical Mean Hight of Back.

Then, by the Rule above, and Chauvenet's Theorem.

Sum of 3 parallel sides of edge and back ÷ 3.

$$\left\{ \begin{array}{l} \left.\begin{array}{l} 80 \\ 80 \\ 80 \end{array}\right\} = \text{Back.} \\ \underline{} \\ 3)240 \\ \overline{80} = \text{Average breadth} \end{array} \right. \qquad \left\{ \begin{array}{l} 10 = \left\{ \begin{array}{l} \text{Vertical Mean} \\ \text{Hight of Back.} \end{array} \right. \\ 100 = \text{Common length.} \\ \underline{} \\ 2)1000 \\ \overline{500} = \text{Area of right sec.} \end{array} \right.$$

Right section × Mean breadth = 500 × 80 . . . . = 40,000 = Volume of *Wedge*.

Smaller end area = 1200 × 100, length . . . . = 120,000 = " " *Prism.*

Solidity of entire prismoid . . . . = 160,000 Cubic Feet.

Proof, by Hights and Widths (SIMPSON).

Hights. Widths.

Larger cross-section . = $50 \times 80 = 4000 = 2\,b.$

Smaller " " . = $30 \times 80 = 2400 = 2\,t.$

Sums of hts. and wids. = $\overline{80} \times \overline{160} = 12800 = 8\,m.$

$$\text{Divisor} = 12)19200$$

$$1600 = \text{Prism. Mean Area.}$$

$$100 = \text{Common length.}$$

Solidity of entire Prismoid (as above) = 160,000 Cubic Feet.

Note.—By HUTTON's *General Rule* we have the same *Solidity* = 160,000 Cubic Feet.

Example 2.—Let us now take the case figured for another purpose, by *Fig.* 14, *Art.* **8.**

Areas.

Large end section . . . . . . = 654 to road-bed only.

Small " " . . . . . = 300 " " "

Difference, or area of back ⎱ = 354
of superposed wedge . . ⎰

Supposing the smaller end, at a distance of 100 feet forward, to be ABKH = 300 in area. While the larger end ABCDEFGHA = 654 area. Common length = 100 feet.

Widths.

$$\text{Then,} \quad \frac{54 + 40}{2} = 47, \text{ Mean width of back.}$$

$$\text{and} \quad \frac{7\cdot532 \times 100 \text{ length}}{2} = 376\cdot6 \quad {\small \text{Right Section.}}$$

$$\frac{354}{47} = 7\cdot532, \text{ Vertical Mean Hight of Back.}$$

9

$$\frac{54 + 40 + 40 = \text{Sum of the three parallel sides}}{3} = 44\tfrac{2}{3} \text{ feet.}$$

$$\textit{Finally,} \begin{cases} 376\cdot6 \times 44\tfrac{2}{3} \quad . \quad . = 16822 = \text{Volume of Wedge.} \\ 300 \quad \times 100 \quad \text{length} = 30000 = \quad `` \quad `` \quad \text{Prism.} \end{cases}$$

$$\left. \begin{array}{l} \textit{Solidity of the whole Prismoid,} \\ \textit{from road-bed to ground line} \end{array} \right\} = 46822 = \text{Cubic feet to road-bed,}$$

or 56,822 to inter-
section of slopes.

Now, roughly computing this example, both by Hights and Widths, and by Roots and Squares, we find for the *Solidity* about the same result, the difference being small in the whole body of earthwork considered.

In like manner, roughly calculating *Figs.* 43 and 44, which have very irregular ground lines, with both end sections in each case *diagrammed upon one figure.* We find that computed by Wedge and Prism, and some other methods, as a proximate test, they *all* coincide within a few cubic yards.

So that this rule for calculating Prismoids of Earthwork by means of a Prism and Wedge, *superposed,* may be accepted as proximately correct in all ordinary * cases, *and it is in practice a very simple one,* as may be noticed in the examples.

Requiring for *data given* merely the areas of the end cross-sections, their distance apart, and their total widths across, horizontally, to ground edges of slopes:—*no matter how irregular the surface may be.*

In all the computations above (as well as in the methods of preceding chapters), so soon as the mean area of an earthwork solid *is ascertained,* it will be found conducive, both to expedition and to accuracy, to resort with it to the table of cubic yards for mean areas (at the end of the book), to obtain cubic yards, *if they should be required in the resulting volume.*

In this connection it may be observed that the transverse area of the under-prism *being always given in the data* (and usually given as that of the smaller cross-section), whilst the distance apart sections is also known, it is better, where cubic yards are desired *in the ultimate solidity,* always to find them from the table in the manner shown by the directions for its use; and the superposed wedge may be also treated in a similar way by computing *its mean area.*

* Where the cross-sections appear to be *unusually distorted,* so as to render doubtful, the application of any ordinary rules, then we must endeavor to sketch an accurate mid-section, and use our First Method of Computation (Chapter II.)—*which never fails* when the data is correct.

29. . . . . . . Although the foregoing rule for the computation of a Prismoid, by Wedge and Prism, *is proximately correct in all ordinary cases*, it has *limits* which must be observed, when exact results are sought.—These limits are: *That the extreme horizontal width of the smaller end section shall always be equal to, or less than, that of the larger end, and never greater, where our rule is used as written above.*

Thus, in all the cases computed in the above examples, the width of smaller end is *less*, except in the figure next preceding, where it is *equal*—but in none of the examples *is it greater*, and hence they are all clearly within the limits of the rule.

In the following figure (*Fig.* 81), however, the horizontal width of the smaller end is, in this unusual case, *greater* than that of the

Fig. 81.

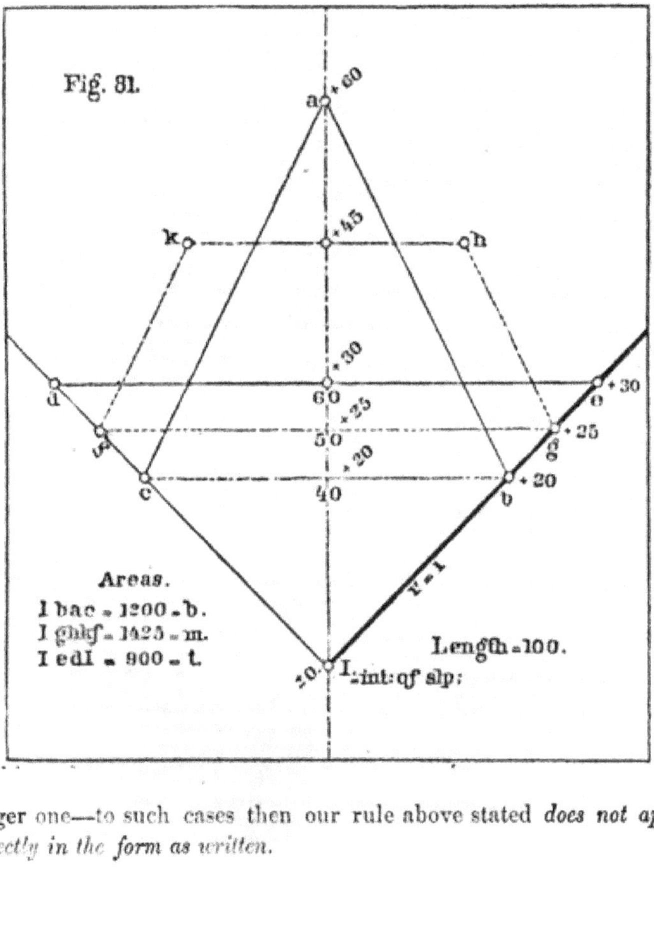

Areas.
I bac = 1200 = b.
I ghkf = 1425 = m.
I edl = 900 = l.

Length = 100.

L.int: qf slp:

larger one—to such cases then our rule above stated *does not apply directly in the form as written.*

A consideration of the figure annexed, where both end sections and the mid-section are diagrammed together, will make the reason evident.

It is simply *this*, that whenever the horizontal top line of the smaller end exceeds in width that of the larger one, or lays *above it* (in a cut), when diagrammed together in one figure, with the diedral angle common to both, *then the smaller end ceases to be the section of a prism, and becomes that of a prismoid.*

But as *a prismoid* is formed of an under prism, with a wedge superposed, we have then in this solid (such as is sectioned in *Fig.* 81) *a prism with two wedges superposed*—the upper one carrying the ground surface of the earthwork solid.

The prism in this case has for its cross-section the portion of the solid *below* the line *c b*, marking the extreme breadth of the larger end section, while the *two* superposed wedges are reversed in position —that in contact with the under prism *having its edge* in the line *c b*, the width of the larger, while that carrying the ground surface *has its edge* in *e d*, the width of the smaller end section; and therefore the wedges are reversed in position, though having the same length in common with the prism, *which underlies both.*

Example 3, Fig. 81.

Data $\begin{cases}$ Cross-section of prism below $c\,b =$ 400.
"   "   smaller end   = 900.
"   "   larger end   = 1200.
Common length of all = 100 feet; other dimensions as in
Fig. 81. $\end{cases}$

(1) *By Prismoidal Formula*—First Method Computation, Chapter II. (Hutton's General Rule)—*which is an accepted standard for accuracy.*

Computation $\begin{cases}$
Smaller end section . . . = 900 = t.
Larger "   "   . . . = 1200 = b.
Mid-section deduced, being
  a mansard figure flat on
  the top = 1425 × 4 . . = 5700 = 4 m.
                        6)7800
                          1300 = Prism. Mean Area.
                           100 = Common length.
Solidity . . . . = 130,000 Cubic Feet. $\end{cases}$

(2) *By Chauvenet's Theorem, and our rule drawn from it.*

Computation.

(1) = *The top wedge (at ground)* = Right
section (40 × 100 ÷ 2 = 2000)
× ⅓ sum of edges = (60 + 40
+ 0 ÷ 3 = 33⅓) . . . . . . = 66,667 C. Feet·

(2) = *The intermediate wedge*, adjoining the
prism (*as in our rule*). Difference
of areas ÷ ½ sum of widths = 500
÷ 50 = 10, Mean Hight of wedge.
Then, by the rule (from Chauvenet), (10 × 100 ÷ 2 = 500) ×
⅓ sum of edges = (60 + 40 + 40
÷ 3 = 46⅓) . . . . = 23,333 " "

(3) = *The prism*, which underlies both =
400 area × 100 length . . . . = 40,000 " "
Totality of this solid, containing two
wedges and one prism = *Solidity* = 130,000 C. Feet.

In examining the solid body terminated by the cross-sections figured
(in *Fig.* 81), it will be found to be bounded *upon every side* by planes,
passed through three common points, so connected that the faces contain *no warped surfaces whatever.*

30. It would appear that in peculiar solids, like that in *Fig.* 81,
we might omit *the prism* entirely, and decompose the body into a
species of double triangular or rhomboidal wedge (with base of back,
and also the edge, common to two triangular wedges superposed, and
inverted with their bases in contact, one on the other), and this
double triangular wedge, with a single pyramid based upon the
smaller end (or in fact on either end), all having a common length,
would form the whole earthwork solid, and simplify the calculation
in such special cases—*if not in all cases of irregular ground.*

Thus, examining the large end I *b a c*, we find it to consist of the backs
of two triangular wedges, joined together at their bases *c b*, and having a common edge at 100 feet forward, equal to *d e*, the top of the
smaller end.

Below this double wedge we find a pyramid whose base is I *e d* I, and
vertex at I, with the common length of 100—the **calculation of
solidity** *is as follows:*

Example 4 (*Fig.* 81).

(1) *The Double* (*Triangular or Rhomboidal*) *Wedge.*

The mean breadth being common both to the upper and lower triangular part of the larger cross-section, then we have, $\dfrac{40 + 60 + 0}{3}$ = 33⅓.

And the whole hight of the double triangular wedge is composed of the hights of the two separate parts = 40 + 20 = 60, forming a Rhomboid.

Then, $\dfrac{60 \times 100}{2}$ = 3000 = Right Section.

And right section = 3000 × ⅓ sum edges = 33⅓ . . . = 100,000

(2) *The Pyramid,* based on smaller end = $\dfrac{900}{3} \times 100$. = 30,000

Solidity of the whole Prismoid . . . . = 130,000

(Being the same as in *Example* 3.)

We might also divide this solid into two wedges and a pyramid by other cutting planes, with the same result. Thus:

Example 5 (*Fig.* 81).

(1) *Upper Wedge,* $\dfrac{40 \times 100}{2}$ = 2000 $\times \left(\dfrac{40 + 60 + 0}{3} \right)$ = 66,667

(2) *Intermed. Wedge,* $\dfrac{30 \times 100}{2}$ = 1500 $\times \left(\dfrac{60 + 40 + 0}{3} \right)$ = 50,000

(3) *Pyramid underlying both* = $\dfrac{400}{3}$ = 133⅓ × 100 length = 13,333

Solidity of the whole Prismoid . . . . = 130,000·

(Being the same as in *Examples* 3 and 4.)

Suppose now upon the smaller end section (*Fig.* 81) we place a triangle of 60 feet base, and 10 feet altitude, the vertex representing the termination of the crest of the ridge coming from the apex of the taller section, and thus augment the area of the lesser end to an equality with the other, or *make each* = 1200 in area—the addition in *Solidity* being *a Pyramid.*

Then, although the end areas are now *equal,* the horizontal widths between the ground edges of the side-slopes *remain unequal,* as before; the big end having least width.

And the computation of this solid is *as follows :*

Example 6 (*Fig.* 81).

By Hutton's General Rule.	*By known Geometrical Solids, governed by Familiar Rules.*

End Areas . . $\begin{cases} = 1200 = t. \\ = 1200 = b. \end{cases}$

m, The mid-section deduced, being a mansard figure, peaked upon the top = 1500 *in area.*

$\dfrac{50 + 30}{2} =$

$40 \times 20 = 800$

$\dfrac{30 \times 5}{2} = 75$

$\triangle$ *of* 25° $= 625$

$\overline{1500}$

$\Bigg\} \times 4 = 6000 = 4\ m.$

$6)\overline{8400}$

$\overline{1400}$ Pris. Mean.

100 Length.

$Sol. = 140,000$ C. Feet.

Pyramid (super-added) base 300.

Then,

$\dfrac{300}{3} \times 100\ .\ .\ .\ .\ .\ .\ . = 10,000$

(1) *Top Wedge* . . . . . $= 66,667$

(2) *Intermediate Wedge* . . . $= 23,333$

(3) *Prism* . . . . $= 40,000$

Solidity in C. Feet . . $= 140,000$

In all the above examples (except *Example* 2), the computation for *solidity* extends from ground surface to intersection of slopes, without regard to the road-bed. But any width of road-bed may be assumed, the volume of the grade prism ascertained, and being *deducted*, will leave the solidity from road-bed to ground all the same, as if it had been specially calculated in that way.

a. . . . . . *Of the Rhomboidal Wedge and Pyramid.*

A close examination of the solid, cross-sectioned in *Fig.* 81, and shown in isometrical projection by *Fig.* 82, will make it evident that beginning with the larger end section, the three cross-sections required by HUTTON's *General Prismoidal Rule* will be a Rhomboid, a Pentagon, and a Triangle, dimensioned as shown in the figures.

And the *solidity* of this body by HUTTON's Rule, as shown in *Example* 3, *Art.* **29** = 130,000 Cubic Feet.

It is also evident, from *Example* 4, of this article, that this computation can be made for *solidity* with the same result (130,000 Cubic Feet), by decomposing the body into a Rhomboidal Wedge and two Pyramids, which may be aggregated and calculated *as one,* so that, as in *Example* 4, this solid can be computed as though it were composed of a single Rhomboidal Wedge, having its edge in the width line of the smaller end section; and of a single Pyramid upon a base equivalent to the latter in area, and its vertex at the foot of the rhomboidal

back which forms the area of the larger cross-section, or one equivalent thereto, and standing (as both end sections do) with the vertices
of one of their vertical angles coincident with the line of intersection
of the side-slopes prolonged.

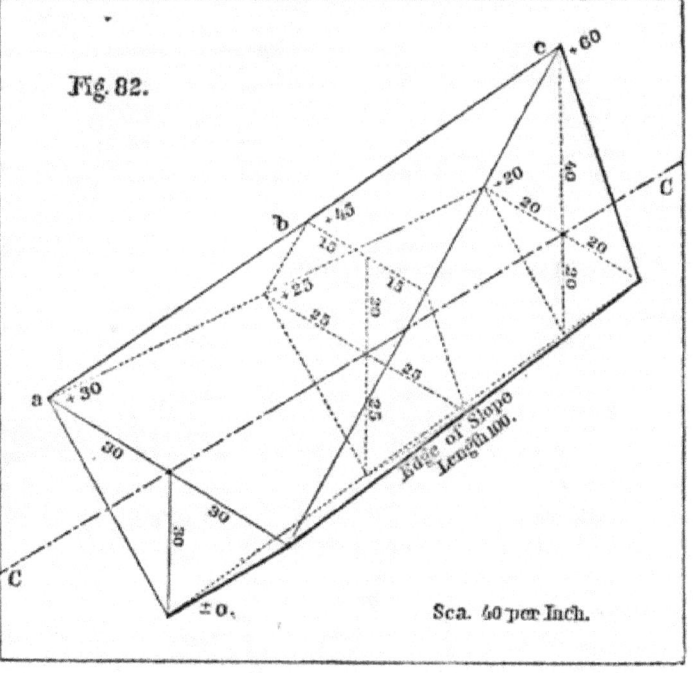

By means of Wedge and Prism, or Wedge and Pyramid (especially
the latter), we have already indicated the process of reaching the volume of an earthwork solid, and we will now continue our examples
until the simple combination of Wedge and Pyramid, in computing
solidity upon the usual earthworks, is fully illustrated.

Although solids resembling *Fig.* 81 in their cross-sections admit of
being easily computed by their own dimensions, either by Wedge,
Prism, and Pyramid, *or by* HUTTON's *General Rule,* which is a standard for volume; nevertheless, as earthwork sections generally present themselves in a somewhat different form, it becomes desirable to
devise a rule which, within a long range, will apply to all earthwork
with uniform slopes, and shall include within its limits the great
majority of cases which come under the notice of the engineer.

Extremely irregular and distorted solids, however, have sometimes to be subjected\ to calculation, which seem almost incommensurable by any fixed rule, and such exceptional cases must be left to independent methods adopted at the time; though it is obvious that any solid may be so sectioned, and divided into limited portions, as to admit of computation by many processes, *without material error.*

b. . . . . *Statement.* In any earthwork solid contained within a diedral angle (formed by the intersection of uniform side-slopes), *however irregular the ground may be,* if the side-slopes continue uniform—and we have *given,* the length *l,* the areas of the cross-sections at the ends A and A', and the slope ratio *r.* We may compute the volume of such solid as a double Triangular, or single Rhomboidal Wedge in combination with a single Pyramid (the latter also usually Rhomboidal but sometimes Triangular).

Process.—Take any pair of irregular cross-sections, judiciously located and measured by the field engineer, so as correctly to define the ground, and of which all the necessary dimensions are known, as well as the distance apart sections.

1. Ascertain the areas of the cross-sections to intersection of side-slopes.

2. Find the proper hight from intersection of slopes, to include one-half the area, also the proper width, and assume this as the base of the back of a double Triangular, or Rhomboidal Wedge in the larger end, and as the edge of the same in the smaller one.

3. Compute from the *larger,* or from *either* end section, a Rhomboidal Wedge, by Chauvenet's Theorem. (See *Example, Art.* **27, a,** paragraph 8.)

4. Then, to the *solidity* of this Rhomboidal Wedge, add that of a Pyramid, based upon the other end section, and having for its altitude the common length, or distance apart sections. (See rule following.)

The sum of the altitudes of the double triangles (joined at their bases) forms the vertical diagonals, or hights of back, of the rhomboidal wedges, while their horizontal·diagonals form the width of back at one end, and of the edge at the other, the angular points of the Rhomboid, *vertically,* being zero. *Either end* may be calculated from, while the other area is the base of a pyramid (Rhomboidal, Triangular, or Irregular), having for altitude the common length *l.* For proof of the work we should always make *both direct and reverse calcu-*

lations, taking either end alternately as the base, and though they will seldom agree *exactly*, owing to the decimals coming in a different order (unless we use a cumbrous number of places); nevertheless, the agreement will be found close enough for a verification of such · work.

To compute the Rhomboidal Wedge and Pyramid in an Earthwork. Adopt either end for *Base*, and call the other *the Top* = *b* and *t*, of former notations.

Present notation:

A = Area of cross-section assumed for *the Base*.
A′ =   "   "   "   "   " *Top*.
l = Common length, or distance apart sections.

These are all the data *required to be given*, the remainder needed are easily computable.

h } Vertical diagonals of the equivalent Rhomboids, into which
h′ }   the end areas are transformed.

w }
w′ } Horizontal diagonals of the same.

Then, by computation:

$$ \left\{ h = 2 \sqrt{\frac{\frac{1}{2} A}{r}};\; h' = 2 \sqrt{\frac{\frac{1}{2} A'}{r}};\; w = \left(\sqrt{\frac{\frac{1}{2} A}{r}} \right) \times 2\,r; \right. $$

$$ w' = \left(\sqrt{\frac{\frac{1}{2} A'}{r}} \right) \times 2\,r. $$

From the foregoing it is evident that $w = h\,r$, and $w' = h'\,r$. Also, when the slopes are 1 to 1, then $h = \sqrt{2\,A}$; if 1½ to 1, $h = \sqrt{\frac{4}{3}\,A}$; and if 2 to 1, $h = \sqrt{A}$. The use of these will often *be convenient.*

RULE.—*Case* 1.—Where width of big end *is equal to, or greater than*, that of small end.

1 (Half product of vertical diagonal of *base*, by distance apart
    sections) × (One-third the sum of horizontal diagonals of
    both ends) = *Solidity of Rhomboidal Wedge;*

$$ \text{or,} \left(\frac{h \times l}{2} \right) \times \left(\frac{w + w'}{3} \right) = S. $$

2 (One-third of area of *top*) $\times$ (Distance apart sections) =
Solidity of Pyramid;

$$\text{or,} \left(\frac{A'}{3}\right) \times l = S.$$

3. Add together the two solidities above (1 and 2) for *the solidity of the entire Prismoid:*—from ground to intersection of slopes, and minus the volume of the grade prism, *gives solidity from road-bed to ground.*

RULE.—*Case* 2.—Where width of big end *is equal to, or less than,* that of small end.

In this case the multiplier for edges (No. 1, Case 1) is to be $\dfrac{(w + w') + (w - w')}{3}$, instead of simply $\dfrac{(w + w')}{3}$. While to the volume produced by the Rule of Case 1—modified in the multiplier as just mentioned—we must *add* a final correction, *as follows:* (Difference of *actual* horizontal widths $\times$ Difference of their hights from intersection of slopes) $\times$ length—this final product, *added* to the volume resulting from *the rule above,* gives the *solidity* for Case 2.

The application of these corrections will be shown hereafter by an example, drawn from the peculiar solid, figured in *Figs.* 81 and 82.

The results produced by these corrections, when *added* to those obtained by the Rule of Case 1, will give the *solidity*, whenever the *actual* width of the smaller end section *does not exceed three times that of the greater one.*

Within these limits the rules and corrections above will apply, and they will be found to cover the great majority of practical cases; but where the end sections are even more distorted, we must then compute by Hutton's General Rule, or by the actual dimensions of the solid, *decomposing it into elementary bodies.*

As the *Prism, Wedge,* and *Pyramid,* are the solid elements from which every great-lined body is composed, *and into which it may be again resolved,* it follows by parity of reasoning (as in the case of the Prismoidal Formula) that for *all* earthwork solids, bounded by planes, the rules of this chapter hold.

c. . . . . . We will now illustrate our method of *Wedge and Pyramid,* by computing the cases of Chapter II., figured from 53 to 64 inclusive, and all originally computed by HUTTON'S *General Rule*—the standard for accuracy.

All of these examples (as indeed is the fact with most others in practice) come under our *Rule and Case* 1—the width of the larger end section being in every instance *greater* than that of the smaller one. (See *Figs.* 53 to 64, *Art.* **18.**

Art. **18.**—*Example, illustrated by Figs.* 53 *to* 55.

$$\text{Given areas to intersection of slopes, etc.} \begin{cases} b = 990 = A \\ t = 500 = A' \\ l = 100 \text{ feet.} \end{cases} \quad \text{Vertical diagonals computed.} \begin{cases} h = 44\cdot50 \\ A' = 31\cdot62 \end{cases} \quad \text{Horizontal diagonals computed.} \begin{cases} w = 44\cdot50 \\ w' = 31\cdot62 \end{cases}$$

The road-bed being 20 feet; the side-slopes 1 to 1 in this case, as in all where $r = 1$; the Rhomboid becomes a square, and the diagonals *equal*.

Direct calculations.

$$\frac{h \times l}{2} \times \frac{w + w'}{2} = S. \text{ of Wedge.}$$

$$\frac{44\cdot50 \times 100}{2} \times \frac{44\cdot50 + 31\cdot62}{3} \quad \therefore = 56,471 = Wedge.$$

$$\frac{A'}{3} \times l = S. \text{ of Pyramid.}$$

$$\frac{500}{3} \times 100 \ldots \ldots \ldots \ldots = 16,667 = Pyramid.$$

$$Total \ldots \ldots \ldots \ldots = 73,138 \quad C. \text{ Feet.}$$

Deduct Grade Prism $\ldots \ldots \ldots = 10,000$

Leaves Solidity of Earthwork $\ldots \ldots = 63,138$

As computed in Art. **18**, Chapter II. $\ldots = 63,170$

$$Difference \ldots \ldots \ldots \ldots = -32$$

Reverse calculations.

$$\frac{31\cdot62 \times 100}{2} \times \frac{31\cdot62 + 44\cdot50}{3} \ldots \ldots = 40,126 = Wedge.$$

$$\frac{990}{3} \times 100 \ldots \ldots \ldots \ldots = 33,000 = Pyramid.$$

$$Total \ldots \ldots \ldots \ldots = 73,126 \quad C. \text{ Feet.}$$

Deduct Grade Prism $\ldots \ldots \ldots = 10,000$

Leaves Solidity of Earthwork $\ldots \ldots = 63,126$

As computed in Art. **18**, Chapter II. $\ldots = 63,170$

$$Difference \ldots \ldots \ldots \ldots = -44$$

The above example represents an earth-cut *upon three-level ground.*

Art. **18.**—*Example, illustrated by Figs.* 56 *to* 58.

This example represents an earth-cut on *five-level ground,* having a
road-bed of 20; slopes of 1 to 1; length 100 feet.

Computed by our Rule, Case 1, *we have.*

Direct calculations.

$$\begin{cases} \text{Wedge} \quad . \quad = 24,306 \\ \text{Pyramid} . \quad . = 14,367 \\ \qquad\qquad\quad 38,673 \\ \text{Deduct G. P.} = 10,000 \\ \textit{Solidity} \quad . = 28,673 \\ \text{By } \textit{Art.} \textbf{18} . = 28,650 \\ \text{Difference.} = \ + 23 \text{ C. Feet.} \end{cases}$$

Reverse calculations.

$$\begin{cases} \text{Wedge} \quad . \quad = 27,254 \\ \text{Pyramid} . \quad . = 11,467 \\ \qquad\qquad\quad 38,721 \\ \text{Deduct G. P.} = 10,000 \\ \textit{Solidity.} \quad . = 28,721 \\ \text{By } \textit{Art.} \textbf{18} . = 28,650 \\ \text{Difference.} = \ + 71 \text{ C. Feet.} \end{cases}$$

Art. **18.**—*Example, illustrated by Figs.* 59 *to* 61.

This example represents an earth-cut on *seven-level ground,* dimensioned as above.

Computed by our Rule, Case 1, *we have :*

Direct calculations.

$$\begin{cases} \text{Wedge} \quad . \quad = 42,048 \\ \text{Pyramid} . \quad . = 21,700 \\ \qquad\qquad\quad 63,748 \\ \text{Deduct G. P.} = 10,000 \\ \textit{Solidity.} \quad . = 53,748 \\ \text{By } \textit{Art.} \textbf{18} . = 53,733 \\ \text{Difference} = \ + 15 \text{ C. Feet.} \end{cases}$$

Reverse calculations.

$$\begin{cases} \text{Wedge} \quad . \quad = 42,935 \\ \text{Pyramid} . \quad . = 20,800 \\ \qquad\qquad\quad 63,735 \\ \text{Deduct G. P.} = 10,000 \\ \textit{Solidity} \quad . = 53,735 \\ \text{By } \textit{Art.} \textbf{18} . = 53,733 \\ \text{Difference} = \ + 2 \text{ C. Feet.} \end{cases}$$

Art. **18.**—*Example, illustrated by Figs.* 62 *to* 64.

This example represents an embankment upon *nine-level ground,*
very rough.   Road-bed 16 feet; side-slopes $1\frac{1}{2}$ to 1; length 100 feet.

Areas *given* $\begin{cases} t = 828\frac{3}{5} = A \\ b = 644\frac{3}{5} = A' \\ l = 100 \text{ feet.} \end{cases}$ Vertical diago- $\begin{cases} h = 33\cdot24 \\ h' = 29\cdot33 \end{cases}$ Horizontal dia- $\begin{cases} w = 49\cdot86 \\ w' = 43\cdot98 \end{cases}$ nals computed. gonals computed.

Direct calculations.

$$\begin{cases} \dfrac{33\cdot24 \times 100}{2} \times \dfrac{49\cdot86 + 43\cdot98}{3}. \quad \dots \quad = 51,987 \text{ Wedge.} \\[2ex] \dfrac{644\cdot67}{3} \times 100 \; \dots \dots \dots \dots \quad = 21,489 \text{ Pyramid.} \\[1ex] \qquad\qquad\qquad\qquad\qquad\qquad\quad \overline{73,476} \\ \text{Deduct Grade Prism.} \dots \dots \dots = \quad 4,267 \\ \qquad\quad \textit{Solidity} \dots \dots \dots \dots \dots = \overline{69,209} \text{ C. Feet.} \\ \text{As computed in } \textit{Art. } \mathbf{18,} \text{ Chapter II.} \dots = 69,200 \\ \qquad\quad \text{Difference} \dots \dots \dots \dots \dots = \overline{\quad + 9} \text{ C. Feet.} \end{cases}$$

Reverse calculations.

$$\begin{cases} \dfrac{29\cdot32 \times 100}{2} \times \dfrac{49\cdot86 + 43\cdot98}{3}. \quad \dots \quad = 45,856 \text{ Wedge.} \\[2ex] \dfrac{828\cdot67}{3} \times 100 \; \dots \dots \dots \dots \quad = 27,622 \text{ Pyramid.} \\[1ex] \qquad\qquad\qquad\qquad\qquad\qquad\quad \overline{73,478} \\ \text{Deduct Grade Prism.} \dots \dots \dots = \quad 4,267 \\ \qquad\quad \textit{Solidity} \dots \dots \dots \dots \dots = \overline{69,211} \text{ C. Feet.} \\ \text{As computed in } \textit{Art. } \mathbf{18,} \text{ Chapter II.} \dots = 69,200 \\ \qquad\quad \text{Difference} \dots \dots \dots \dots \dots = \overline{\quad + 11} \text{ C. Feet.} \end{cases}$$

d. . . . . . We have thus compared the whole four of the examples illustrated in Chapter II., and all computed by HUTTON's *General Rule.* These we find to agree with the calculations by Wedge and Pyramid, in every instance within a few cubic feet, and had the decimals (into which all these computations run) been carried further, the agreement would probably have been closer.

We will now compute by *Wedge and Pyramid* the example of a heavy embankment, taken from Warner's Earthwork, *Art.* 86.

"Prismoid.   First end-hight — 28·7 ; second end-hight — 14·5 ; surface-slope 15° ; side-slope 1½ to 1 ; road-bed 24 feet."

$$\begin{aligned} &\textit{Data computed}\begin{cases} b = 2411 = A \\ t = 907 = A' \\ l = 100 \text{ feet.} \end{cases} \begin{array}{l}\text{Vertical diago-} \\ \text{nals computed.}\end{array}\begin{cases} h = 56\cdot70 \\ h' = 34\cdot78 \end{cases} \begin{array}{l}\text{Horizontal dia-} \\ \text{gonals computed.}\end{array}\begin{cases} w = 85\cdot05 \\ w' = 52\cdot17 \end{cases} \\ &\text{to intersection of} \\ &\text{slopes, etc.} \end{aligned}$$

Direct calculations.

$$\frac{56{\cdot}70 \times 100}{2} \times \frac{85{\cdot}05 + 52{\cdot}17}{3} \quad \ldots \ldots = \overset{\text{C. Feet.}}{129{,}673} \text{ Wedge.}$$

$$\frac{907}{3} \times 100 \ldots \ldots \ldots \ldots \ldots = \quad 30{,}233 \text{ Pyramid.}$$

$$\overline{159{,}906}$$

For Cubic Yards ÷ 27. . . . . =   5,923

Deduct volume of Grade Prism . . . . . =   356

      Solidity. . . . . =   5,567 C. Yards.

By Hutton's General Rule . . . . . . =   5,566

      Difference . . . . =   + 1 C. Yard.

Reverse calculations.

$$\frac{34{\cdot}78 \times 100}{2} \times \frac{52{\cdot}17 + 85{\cdot}05}{3} \quad \ldots \ldots = \overset{\text{C. Feet.}}{79{,}542} \text{ Wedge.}$$

$$\frac{2411}{3} \times 100 \ldots \ldots \ldots \ldots \ldots = 80{,}367 \text{ Pyramid.}$$

$$\overline{159{,}909}$$

For Cubic Yards ÷ 27. . . . . =   5,923

Deduct volume of Grade Prism . . . . . =   356

      Solidity. . . . . =   5,567

By Hutton's General Rule . . . . . . =   5,566

      Difference . . . . =   + 1 C. Yard.

Mr. Warner (in *Art.* 86 quoted) makes the volume here computed = 5562 *Cubic Yards.*

e. . . . . . All of the above examples come under Case 1, of our Rule, as ordinary earthwork sections *usually do.* But we will now compute a single example by Case 2—where the width of the greater end *is less* than that of the smaller one. This condition will be found in the solid figured in *Figs.* 81 and 82.

In this example, illustrative of the rule in Case 2, the corrections therein named have been duly embodied.

Example of Case 2 (Fig. 81).

$$\frac{48{\cdot}98 \times 100}{2} \times \frac{48{\cdot}98 + 42{\cdot}42 + 6{\cdot}56}{3} \quad . \quad . = 80{,}000 \text{ Wedge.}$$

$$= \frac{h \times l}{2} \times \frac{(w + w') + (w - w')}{3}$$

$$\frac{900}{3} \times 100 \quad . \quad . \quad . \quad . \quad . \quad . \quad . \quad . \quad . \quad . \quad . = 30{,}000 \text{ Pyramid.}$$

$$= \frac{A'}{3} \times l. \qquad\qquad\qquad 110{,}000$$

Final correction, $10 \times 10 \times 20 \times 100$ . . = $\underline{20{,}000}$

Solidity . . . . = $\overline{130{,}000}$ C. Feet.

The same as computed before . . . . . = 130,000

It would appear, then, from the discussion in this chapter, the examples given, and the simplicity and conciseness of the rules for computing earthworks, by means of the *Prism*, *Wedge*, and *Pyramid*, that they deserve to rank amongst the best employed for the purpose.

---

* Although this solid (*Figs.* 81 and 82) is bounded on all sides by plane surfaces, and is composed simply of a Rhomboidal Wedge, superposed upon a Pyramid—very few of the Rules or Tables, of the numerous writers on Earthwork, furnish means for computing its *solidity*—which can only be readily ascertained by HUTTON's General Rule, or by decomposition into elementary solids, of which the rules for volume have been long established.

CHAPTER VI.

31. ..... The late Professor W. M. Gillespie, of Union College,
Schenectady, N. Y., was an able teacher of Civil Engineering, and a
sound practical writer on that and cognate subjects, as may witness
his—Roads and Railroads (1847), 10 editions; Land Surveying
(1855), 8 editions; Higher Surveying, etc. (1870), *posthumous*, 1
edition; and numerous valuable papers, read before the American
Scientific Association, or printed in scientific journals.

In 1847 he published his first edition of Roads and Railroads, and,
as an appendix to it, in about 25 pages, he gave a practical summary
of various methods of computing Excavation and Embankment,
accompanied by valuable corrections and suggestions, which were
together so explicit and so well grounded that this Appendix has
become the basis of several works upon the subject, whose authors,
without much acknowledgment (often without any), have freely
availed themselves of Professor Gillespie's labors.

His work on Roads and Railroads, well printed and cheaply published, has had a great circulation; it has already filled 10 editions,
and is probably better known in the offices of engineers, all over this
country, than any other similar book. In the Appendix, on Excavation and Embankment, Professor Gillespie recognizes "*four usual
methods of calculation.*"

1. Calculation by Averaging End Areas (*or Arithmetical Average*).
2. " " Middle Areas.
3. " " Prismoidal Formula.
4. " " Mean Proportionals (*or Geometrical Average*).

And we will now proceed to give his views substantially, but not
literally, upon these *four rules*, which he found in use when he took
up this subject in 1847, *and which, indeed, had long before been known,*
—*as follows:*

1st. *Arithmetical Average.*—This consists simply in adding together the areas of any two adjacent cross-sections, taking half their sum for a mean area, and multiplying it by the length of the station, or distance apart sections,—*to find the Solidity.*

As generally used by engineers, instead of adding the end areas, halving their sum, etc., *they* employ the sum of the two, *or double areas,* and merely double one of the divisors in working for Cubic Yards, *as follows:*

Engineers' Rule.

Take the sum of the areas of any two adjacent cross-sections, multiply these *double areas* by the length (which, if a full station of 100 feet, is done mentally, or by removing the decimal point two places to the right). Divide by 6 and by 9, and *the last quotient gives the volume in Cubic Yards.*

This Rule has been *by far* the most used of any other in our country ;—with tables of Cubic Yards, for double areas, it is very expeditious, and has found numerous advocates amongst engineers on account of its simplicity and convenience; it usually gives a result *in excess* of the truth, and where the disparity of areas is great, *very much in excess;* even this well-known error has found commendatory advocates, on the ground that it is like the merchant giving good measure to the customer, and that this excess in quantity being well understood, would be compensated for by a reduced price, whenever the work was executed by contract—*but these arguments are clearly unsound.*

Professor Gillespie has, however, indicated a simple correction, by means of which the result of a computation, by Arithmetical Average can be reduced *to the truth.*

Thus, let

d = Difference of centre hights, supposing all the cross-sections to be reduced to an equivalent *level* top.

$s*$ = Ratio of the side-slopes (*or cot. of angle*) s to 1.

l = Length of the cut or fill between sections.

---

* Engineers and writers have pretty generally, of late years, agreed to designate the ratio of side-slopes as r (and this we have usually employed), while the symbol s is confined to slopes of ground, or *surface slopes,* but in the present case Professor Gillespie's *notation* is adhered to.

Then, $\dfrac{s\,d^2\,l}{6}$ is the proper correction for the results of Arithmetical Average, which correction, if computed for each mass so calculated, and then *deducted* therefrom, will give *the true solidity*—the same precisely as if calculated direct by the Prismoidal Formula itself.

The chief example computed by Professor Gillespie under the several heads of his subject, has the same data in all, as shown by the first four columns of the following Tables—the cross-sections in all cases being assumed to be equivalent level trapezoids by him.

1. *Arithmetical Average.*

Table 1, computed in illustration of the corrections proposed, including an entire section of a supposed railroad, 4219 feet in length.

1. Road-bed 50; side-slopes of excavation 1½ to 1; of embankment 2 to 1.

Sta.	Distance in feet.	Cut. + in feet.	Fill. — in ft.	End Areas, or Cross-secs. Sq. Ft.	Excavation. C. Feet. Computed by Arith. Average.	Embankment. C. Feet.	CORRECTIONS. By Formula $\frac{s\,d^2\,l}{6}$		Amounts in Cubic Feet. deductive.	Corrected quantities, agreeing with the Prismoidal Formula. Excavation. C. Feet.	Embkt. Cubic Feet.
1		○		○							
2	561	18		1386	388,773		$1\frac{1}{2}\times 18^2\times 561$ over 6		45,411	343,352	
3	858	20		1600	1,280,994		$1\frac{1}{2}\times 2^2\times 858$ over 6		858	1,280,136	
4	825	○	○	○	660,000		$1\frac{1}{2}\times 20^2\times 825$ over 6		82,500	577,500	
5	820		19	1672	685,520		$2\times 19^2\times 820$ over 6		98,673		586,847
6	825		8	528	907,500		$2\times 11^2\times 825$ over 6		33,275		874,225
7	330	○	○		87,120		$2\times 8^2\times 330$ over 6		7,040		80,080
	4219	38	27	+2986 −2200	2,329,767	1,680,140	6	125,794	138,988	2,200,963	1,541,152

From this Table it will be perceived that the error of the process of Arithmetical Average, in this example, amounts in Excavation to 6 per cent., and in Embankment to 9 per cent., *above the true solidity*.

2d. *Calculation by the **Middle Areas**.*—The second method of calculation is to deduce *the middle areas* (commonly called *mid-sections*) of each Prismoidal mass, from the middle hight, or Arithmetical Mean of the extreme hights of the solid, and multiply the middle area thus found by length for volume. The results thus obtained are *too small;* their *deficiency* being equal to just *half the excess* of the first method.

Here the corrective formula is, $\dfrac{s\,d^2\,l}{12}$; and corrections thus calculated being *added* to the results obtained, by the process of middle areas, would make them coincide *with the true volume given by the Prismoidal Formula*.

2. Middle Areas.

Table 2, *computed and corrected in illustration of the above,* including an entire section of a supposed railroad = 4219 feet in length.

2. Road-bed 50; side-slopes of excavation 1½ to 1; of embankment 2 to 1.

Sta.	Distance in feet.	Cut. + in feet.	Fill. - in ft.	Middle Areas. Sq. Ft.	Computed by Middle Areas. Excavation.	Embankment. Cubic Feet.	CORRECTIONS. By Formula $\frac{s\,d^2\,l}{12}$	Amounts in Cubic Feet, additive. Ex.	Em.	Corrected quantities, agreeing with the Prismoidal Formula. Excavation. C. Feet.	Embankment. C. Feet.
1 2	561	18 ◯		571·5	329,641		$1\frac{1}{2} \times 18^2 \times 561$ / 12	22,721		343,332	
3	858	20		1491·5	1,279,707		$1\frac{1}{2} \times 2^0 \times 858$ / 12	429		1,280,136	
4	825	◯	◯	650	536,250		$1\frac{1}{2} \times 2^0 \times 825$ / 12	41,250		577,500	
5	820		19	655·5		537,510	$2 \times 19^2 \times 820$ / 12		49,337		586,847
6	825		8	1039·5		857,587	$2 \times 11^3 \times 825$ / 12		16,638		874,225
7	330		◯	232		76,560	$2 \times 8^3 \times 330$ / 12		3,520		80,080
	4219	38	27	+2713·0 −1927·0	2,136,568	1,471,557	12	64,400	69,495	2,200,968	1,541,152

From the above Table it will be perceived that this process of *Middle Areas* is a closer one than that of Arithmetical Average; but being *in deficiency*, while the former was *in excess*, the difference in this case, from *the true solidity*, being about 3 per cent. *less* in Excavation, and about 4 per cent. *less* in Embankment.

3d. *Calculation by the Prismoidal Formula.*—The mass of which the volume is demanded *is a true Prismoid*, and its contents will therefore be given by the well-known Prismoidal Formula.

$$\frac{b + 4m + t}{6} \times \text{length} = \textit{Volume.}$$

$$\text{Where,} \begin{cases} b = \text{Area of Base.} \\ m = \text{Mid-section.} \\ t = \text{Area of top.} \end{cases}$$

Retaining the same data for the example as has been used in the preceding tabulations, and will be continued throughout this discussion, we refer to the following Table (3), where the results obtained from the data given, by means of the Prismoidal Formula, are properly tabulated.

3. *Prismoidal Formula.*

Table 3, in illustration of the computation by it. Including an entire section of a supposed railroad = 4219 feet in length.

3. Road-bed 50; side-slopes of excavation 1½ to 1; of embankment 2 to 1.

Sta.	Distance in feet.	Cut. +	Fill. —	End Areas. Sq. Ft.	Middle Areas. Sq. Ft.	QUANTITIES. Excavation. C. Feet.	QUANTITIES. Embankment. C. Feet.
1		O		O			
2	561	18		+1386	+ 571·5	343,332	
3	858	20		+1600	+1491·5	1,280,136	
4	825	O		O	+ 650	577,500	
5	820		19	—1672	— 655·5		586,847
6	825		8	— 528	—1039·5		874,525
7	330		O		— 232		80,080
	4219	+38	—27	+2986	+2714	2,200,968	1,541,452
				—2100	—1927		

This Table 3, computed by the Prismoidal Formula itself, *is the standard* for all the others, and gives *the true solidities* in the section of railroad under consideration.

4th. *Calculation by Mean Proportionals* (or Geometrical Average). —Professor Gillespie says a fourth method, called that of *" Mean Proportionals,"* is sometimes, though very improperly, employed.

He gives the following rule for Mean Proportionals.

    Rule.—Add together the areas of the two ends, and a Mean Proportional between them (found by extracting the Square Root of their product); multiply the sum of these three areas by the length of the Frustum, and divide the product by three.*

As used by engineers, in working for Cubic Yards as the result, this rule takes a somewhat different shape, *as follows:*

    Rule.—Multiply the sum of the end areas, and the Square Root of their product, by the distance apart, and divide *this final product* by 9 and by 9.

---

* This is, *substantially,* Euclid's Rule for the Frustum of a Pyramid; Davies' Legendre, VII. 18.

The result is always much less than the truth (supposing the areas taken between ground line and road-bed), for it treats as Pyramids, or thirds of Prisms, the wedge-shaped pieces which are really halves of Prisms, and is farthest from the truth when one of the areas = 0.* *So far the Professor.*

And this is *all* correct *when* the cross-sections are limited between road-bed and ground surface; but if they are extended to the intersection of the side-slopes, or edge of the diedral angle containing the earthwork solid, *an entirely different state of affairs takes place,* for if the road-bed be imagined to be gradually narrowed, so that eventually it vanishes at the intersection of the side-slopes; then, at that point, both Pyramid and Prismoid *coincide,* or become *equivalent,* whilst their rules become *correlative* (or mutually interchangeable), and *either* may be used with the same results in point of *solidity;* and this is also the case with the "*Equivalent Level Hights,*" much used by engineers since the publication of Sir John Macneill's work (London, 1833), but likewise condemned by Professor Gillespie, rather hastily as it seems to the writer, and hardly upon sufficient grounds.

It seems singular that this able Professor should have overlooked the facts mentioned above, as he was well acquainted with the method of continuing calculations to junction of side-slopes, *including* the Grade Prism in the earlier stages of the computation, but *rejecting* it at the close (as may be seen in his paper on Warped Solids (1859)).

Now, so long as the cross-section of the earthwork remains *trapezoidal* in figure, the strictures of Professor Gillespie upon this rule (commonly called the Geometrical Average) *are undoubtedly correct;* but whenever the cross-section becomes triangular *they fail entirely,* as also does his similar censure on "*Equivalent Level Hights.*"

In evidence of this, we have tabulated (for ourselves) the same general example as heretofore given—both for the Geometrical

---

* Now, taking a case of precisely this kind (only continued to intersection of slopes) —hight at one end 34·5, at the other 0, with road-bed of 30 feet, slopes of 2 to 1, a length of 66 feet, and level on the top.

If we compute this solid, either prismoidally, or by the usual rule for wedges, we have for its volume 3205 Cubic Yards in round numbers.

And if we compute it by Baker's Rule (who treats such cases as Frusta of Pyramids, *but with the important addition of the Grade Prism*), we find the resulting volume to be the same to the nearest Cubic Yard.

For this *pyramidal rule* see Baker's Earthwork, London, 1848, whose rule is similar to that of Bidder and others, which have always been accepted as *correct* by English engineers, *and most certainly they are so.*

Average, and for the Equivalent Level Hights, merely carrying the areas to the intersection of the side-slopes, in both cases, including at *first* the Grade Prism, but *excluding* it after—as a common quantity.

32. . . . . ∴ By these Tables we find the solidity of Gillespie's example to be *precisely the same* as computed by him with the Prismoidal Formula (Table 3 above), and which he has very properly adopted *as the correct standard for all.*

4. *Mean Proportionals (or Geometrical Average).*

Table 4, in illustration of computation by them, including an entire section of a supposed railroad = 4219 feet in length.

4. Road-bed 50; side-slopes of excavation $1\frac{1}{2}$ to 1; of embankment 2 to 1.

Sta.	Distance in feet.	To the Road-bed. Cut. +	To the Road-bed. Fill. —	To intersection of slopes. Cut. +	To intersection of slopes. Fill. —	End Areas to intersection of slopes. Sq. Feet.	End Areas to intersection of slopes. Sq. Feet.	Geometrical Mean Area. Sq. Feet.	Quantities agreeing with those of the Prismoidal Formula. Excava. Cub. Feet.	Quantities agreeing with those of the Prismoidal Formula. Embank. Cub. Feet.
1		○		$16\frac{2}{3}$		416·666				
2	561	18		$34\frac{2}{3}$		1802·666		+ 816·666	343,332	
3	858	20		$36\frac{2}{3}$		2016·666		+ 1906·666	1,280,136	
4	825	○	○	$16\frac{2}{3}$	$12\frac{1}{2}$	416·666	312·5	+ 916·333	577,500	
5	820		19		$31\frac{1}{2}$		1984·5	— 787·5		586,847
6	825		8		$20\frac{1}{2}$		840·5	— 1291·5		874,225
7	330		○		$12\frac{1}{2}$		312·5	— 512·5		80,080
	4219	+38	—27	+ $104\frac{2}{3}$	— 77	+ 4652·664	— 3450·0	+ 3639·665 — 2591·5	2,200,968	1,541,152

In this Table the Grade Prism is *included* at first, and *excluded* afterwards. Its sectional area is *as follows:*

Grade Prism of Cut = 416·666 Square Feet.
" " Bank = 312·5 " "

To be multiplied for volume by length of mass to which it belongs.
Altitudes of the Grade Prism in the Cut = $16\frac{2}{3}$ feet; on Bank = $12\frac{1}{2}$ feet.

In computing quantities by **Geometrical Average**, the following generalization has occurred to the writer, which indeed *may possibly be* a germ from which the Prismoidal Formula might have sprung—since both the Arithmetical and Geometrical Means were known in the days of Euclid (200 B. C.), while the original Prismoidal Formula (so far as we know) was devised by Simpson, as late as A. D. 1750.

Thus,

$$\frac{\text{Double the sum of End Areas} + \text{Double Geom. Mean}}{6} \times h = Solidity.$$

Let

$$\begin{cases} A = \text{Sum of End Areas.} \\ B = \text{Geometrical Mean.} \end{cases} \text{Then the above} \begin{cases} \dfrac{2\,A + 2\,B}{6} \times h = S. \end{cases}$$

Or, in its lowest terms, $\dfrac{A + B}{3} \times h = S$, which is *the Geometrical*

Average; or, *in substance,* Euclid's Rule for the Frustum of a Pyramid; and by the aid of *the Grade Prism* strictly applicable to earthworks of a general triangular section in ordinary cases.

5 *Equivalent Level Hights.*

Table 5, in illustration of computation by them.

5. Road-bed 50; side-slopes of excavation $1\frac{1}{2}$ to 1; of embankment 2 to 1.

Sta.	Distance in ft.	To the Road-bed.		To intersection of slopes.		End Areas to intersection of slopes.		Mid. hts. to intersection of sl'pes.	Mid-sections, or areas to the intersection of slopes.		Quantities agreeing with those of the Prismoidal Formula.	
		Cut. +	Fill. −	Cut. +	Fill. −	Cut. +	Fill. −	Feet.	Sq. Feet.	Sq. Feet	Excava. C. Feet.	Embkt. C. Feet.
1		0		16⅔		416·666						
2	561	18		34⅔		1802·666		+ 25·666	988·166		343,332	
3	858	20		36⅔		2016·666		+ 35·666	1908·166		1,280,136	
4	825	0	0	16⅔	12½	416·666	312·5	+ 26·666	1066·666		577,500	
5	820		19		31½		1984·5	− 22·000		968·0		586,847
6	825		8		20½		840·5	− 26·000		1352·0		874,225
7	330		0		12½		312·5	− 16·500		544·3		80,080
	4219	+38	−27	+ 104⅔	−77	+ 4652·654	− 3450·000	+ 87·998 − 64·500	+ 3962·998	− 2864·5	2,200,968	1,541,152

In this Table the Grade Prism is *included* in the earlier operations, and *excluded* in the later ones. Its sectional area is *as follows:*

Grade Prism of Cut = 416·66 Square Feet.

"   "   Fill = 312·50   "   "

To be multiplied for volume by the length of mass to which it belongs.

Altitudes of the Grade Prism in the Cut = $16\frac{2}{3}$ feet; on Bank = $12\frac{1}{2}$ feet.

33. From the preceding discussion in the present chapter we are justified in declaring that all the following rules and formulas (*detailed above*) are *equivalent* in their results for *volume*—when properly corrected and appropriately used; and that they all give *the same solidity in the end* as No. 3 does, which is the standard for ALL.

1. Arithmetical Average to Road-bed (with correction).
2. Middle Areas to Road-bed (with correction).
3. Prismoidal Formula (*the standard for all*) to Road-bed, or to the intersection of slopes—*either.*
4. Geometrical Average to intersection of slopes.
5. Equivalent Level Hights to intersection of slopes.

All these are fully described above, and the tabular statements bearing the same number show in each case the results of the calculations for volume, agreeing uniformly with the computations for solidity, *made by means of the Prismoidal Formula.*

In concluding his notices of the method of computing the contents of earthworks, by means of the Prismoidal Formula, Professor Gillespie gives some special rules, transformed from it, which are doubtless valuable in certain cases, but do not appear to be of general application; he also gives formulas for a series of equal distances apart stations, such as are usually found in the location of railroads.

These are intended to be applied to *a central core*, or body of the work, based upon the road-bed, to be calculated by itself, and then *the slopes*, to be computed separately or together, and added in with the core, so as to form finally *the volume of the whole prismoidal mass.*

This idea of separating the core or body from the slopes, calculating them independently, and adding them together, seems to have occurred to a great many engineers,* and forms the theme of nearly a dozen books on the subject of Earthwork Measurements—*here or abroad.*

Indeed, the very first special work on the mensuration of earthworks, which was published in this country—that of E. F. Johnson, C. E. (New York, 1840), adopted this system, and furnished a series of Tables to facilitate its operation;—it was, however, briefly explained before, in Lieut.-Col. Long's valuable Railroad Manual (Baltimore, 1828), which was the first to treat the subject in this country, *and was, in fact, the pioneer of technical railroad literature in the* UNITED STATES.

Nevertheless, the method of *Core and Slopes* has never come into general use, though often revived from time to time by new writers, apparently unacquainted with the literature of this subject.

---

* Amongst others, it is the method of Bidder, who followed Macneill in the earlier days of English railroads.

34. . . . . . . *Comparison of Gillespie's Main Example and the Method of Roots and Squares.*

Professor Gillespie's chief example, of a heavy Cut and Fill, forming an entire section of railroad, 4219 feet long, must by this time be so familiar to engineers, and others, in consequence of the extensive circulation of his Manual of Roads and Railroads, since its original publication in 1847, that we have selected it as the most suitable, or at least *the best known,*[*] for the purpose of comparison with our Third Method of Computation—*that by Roots and Squares.*

We therefore give a Table No. 6 (below), which contains in the first 5 columns *the data* given by Professor Gillespie, and in the last 6 *the results* of the computation by Roots and Squares, which will be found to agree *exactly* with those obtained above, by means of the Prismoidal Formula—*accepted as being a correct standard for comparison.*

6. *Comparison of Example, with Roots and Squares.*

Including (as before) an entire section of a supposed railroad = 4219 feet in length.

6. Road-bed 50; side-slopes of excavation 1½ to 1; of embankment 2 to 1.

Sta.	Distance in feet.	End Areas in Sq. Ft. Out + Fill —	Centre Hights in feet. Cut + —	End Areas increased by Grade Triangle. Sq. Feet.	Square Roots of End Areas. Feet.	Sums of Square Roots. Feet.	Squares of sums, or 4 times the mid-section. Feet.	Quantities agreeing with those given by the Prismoidal Formula. C. Feet.	C. Feet.
1		O	O	+ 416½	+ 20·42				
2	561	+1386	18	+1802½	+ 42·46	62·88	3954	313,232	
3	858	+1600	20	+2016½	+ 44·91	87·37	7634	1,283,130	
4	825	O	O O	{ + 416½ / − 312½	+ 20·42 / − 17·68	65·33	4268	577,500	
5	829	−1672	19	−1984½	− 44·55	− 62·23	− 3872		586,847
6	825	− 528	8	− 840½	− 28·99	− 73·54	− 5498		874,225
7	330	O	O	− 312½	− 17·68	− 40·67	− 2178		80,080
	4219	+2986 / −2200	+38 / −27	+4652¾ / −3450	+ 128·21 / − 108·90	215·58 / − 182·44	15856 / − 11458	2,200,968	1,541,152

In the above Table (as in the others), the cross-sections—in the data given—being level trapezoids from ground to road-bed, we neces-

* Besides, this example, originated by F. W. Simms, C. E. (London, 1836), has been before the public *for many years,* having been first published in our country in Alexander's edition of Simms on Levelling (Baltimore, 1837); from which, or the original, it was copied by Professor Gillespie. In the work above mentioned, Mr. Alexander gives every detail of the computation of this example, by the Prismoidal Formula, *at great length,* and so indeed does Simms.

sarily *add* in this mode of computation (to intersection of slopes) the Grade Triangle, and *deduct* it again near the close of the operation.

Road-bed 50; side-slopes of excavation $= 1\frac{1}{2}$ to 1; of embankment $= 2$ to 1.

Grade Triangle of Cut, area $= 416\frac{2}{3}$ Sq. Ft. — altitude $= 16\frac{2}{3}$ Feet.
　　"　　"　　" Fill, " $= 312\frac{1}{2}$ " " — " $= 12\frac{1}{2}$ "

Where the distances apart stations *are uniform in length and even in number*, the method of Roots and Squares enables us to employ a very simple modification of Simpson's Multipliers, as has been already shown in Chapter IV., so as to compute with ease and expedition an entire cut or fill, *at a single operation*, or one station only, *at pleasure*.

CHAPTER VII.

35. ...... Preliminary, and often hasty estimates of earthworks,
are constantly required by engineers prior to deciding upon railroad
routes, or their modifications, and indeed are *generally* necessary in
determining the relative merits of engineering lines—(amongst which
there are always *alternatives*)—since few can undertake to settle properly any important questions relating to their comparative value,
without some serious consideration, for which the Preliminary Estimates, on various lines surveyed, supply a proximate foundation, by
aiding without controlling the judgment of the engineer.

Exploring Lines, preparatory to the final location of a railway, are
indispensable, and in a difficult country may extend to *tenfold* the
length of *the final line,* while the time allowed to engineers being usually
extremely short, the estimates of quantities on these Preliminary Surveys are necessarily hasty, and consequently *imperfect*—but nevertheless demand rapidity in execution, *however made.*

For this there seems to be no remedy; all we can do is to endeavor
to point out a method for hasty estimates, *more correct and more expeditious* than those usually employed, and to this we shall confine ourselves in the present chapter.

Exploring lines are usually traced with stations *at double distance,*
or 200 feet apart—and, indeed, sometimes on plain ground the distance apart stations has been stretched (to save time) as far as 400
or 600 feet;—and as this last distance is about the longest range
which gives *distinct vision* for the Engineer Levels in use in this
country, it ought rarely to be exceeded, as a general rule; while at
least, the distance of 200 feet apart stations, *or double distance of loca-*

156

tion, furnishes good information of the ground, and also enables the exploring party to proceed rapidly enough to gain an adequate knowledge of the country, *without much loss of time.*

Nevertheless, the rules we suggest will apply to any *uniform distance* apart stations of exploring line, which may be deemed advisable by the engineer in charge: but the longer the distance between stations, the less accurate will be the estimate *in general.*

We propose to apply Simpson's celebrated rule for cubature (the accuracy of which is well known) to Preliminary or Hasty Estimates, *taking as data* the centre hights and surface slopes *alone;* the former to the nearest foot of hight or depth, from ground to intersection of side-slopes, and the latter to the nearest 5° of average ground slope across the line, leaving special cases to be dealt with by the engineer, according to rules of his own.

We have provided proximate tables (very nearly correct) to facilitate these hasty operations, and would also suggest that, in all cases of Preliminary Estimates, the resulting quantities of earthwork should be augmented *ten per cent.:*—this addition will give *full quantities*, and has been shown by long experience to be *ample* to meet the usual contingencies which always arise in the construction, and cannot be foreseen, and of which, in fact, it must be confessed, the engineer in charge (often unknown to himself) *almost invariably takes the most favorable view,* and hence the greater necessity exists for some appropriate allowance beyond the net result of the calculations.

Simpson's Rule for Cubature, using cross-sections instead of ordinates (as we have before shown), is *as follows:*

$$\frac{A + 4B + 2C}{3} \times D = Solidity.$$

(Sometimes 2 D, and 6 for divisor, are used, and are *equivalent.*)

A = Sum of extreme end ordinates, or sections.
B = Sum of cross-sections standing on *even* numbers.
C = Sum of " " " " *odd* numbers.
D = The common interval, or distance apart sections.

Simpson's rule above is limited to an even number of equal spaces.

And it must be observed that in its application it is always best to prepare a rough profile of the line run, and under the regular numbers to pencil forward, from the beginning of the cut or fill to be computed, the series of numbers 1, 2, 3, 4, etc. No. 1 always standing at the place of beginning; it is this series of numbers, so arranged, which are referred to in the rule above as *even and odd*.

By this rule it is best to compute *entire and separately* each cut and each fill encountered by the line; and if the whole number of *equal* intervals or stations, in any cut or fill, should be *an odd number*, then one station of the common length, at beginning or end (or indeed any where deemed most suitable), should be struck off temporarily, and reserved for separate calculation; while the body of the work thus reduced, *to an even number of common intervals*, comes directly within the rule, and can be calculated as a whole, while the detached station, computed by itself, may be added in near the close of the operation.

It will always be found briefer and better in using this and similar rules, to aim first at finding a *General Mean Area*, which, multiplied by the proper length or distance, will give *the solidity;* but it is still better, having the General Mean Area in square feet, to use our Table at the end when the result is desired *in Cubic Yards.*

36. . . . . . . Instead of employing Simpson's Formula, as it stands above, it will be often more convenient to use the multipliers which represent it—these are known as *Simpson's Multipliers,* * and are *as follows:*

For *two* equal intervals, apart sections, *Mults.* =   1, 4, 1.   { Divisors 6; quotient, Mean Areas; factors for length = *double interval,*

" *four* "   "   "   "   "   =   1, 4, 2, 4, 1.   { Divisors 3; quotient. Mean Areas; factors for length = *single interval.*
" *six* "   "   "   "   "   =   1, 4, 2, 4, 2, 4, 1.
" *eight* "   "   "   "   "   =   1, 4, 2, 4, 2, 4, 2, 4, 1.
" *ten* "   "   "   "   "   = 1, 4, 2, 4, 2, 4, 2, 4, 2, 4, 1.

The first set of multipliers, their divisors, and factors for length, are clearly *those of the Prismoidal Formula,* which evidently forms the basis of this famous rule.

Indeed, it is easy to show by diagrams how this rule may probably have been formed, by the eminent mathematician, with whom it originated, about the year 1750; and also how intimately it appears to be connected *with the Prismoidal Formula.*

---

* Rankine's Useful Rules and Tables, 2d edition, London, 1867, page 64.

See *Figs.* 77 and 78, *following.*

Suppose Figs. 77 and 78 to represent front views of four planes, A, B, C, D, or of four solids with a thickness of *unity*, all standing on the level base line EF, and that their respective ordinates, or cross-sections (correlative in Simpson's Rule for Cubature), are *dimensioned* as marked upon the figures.

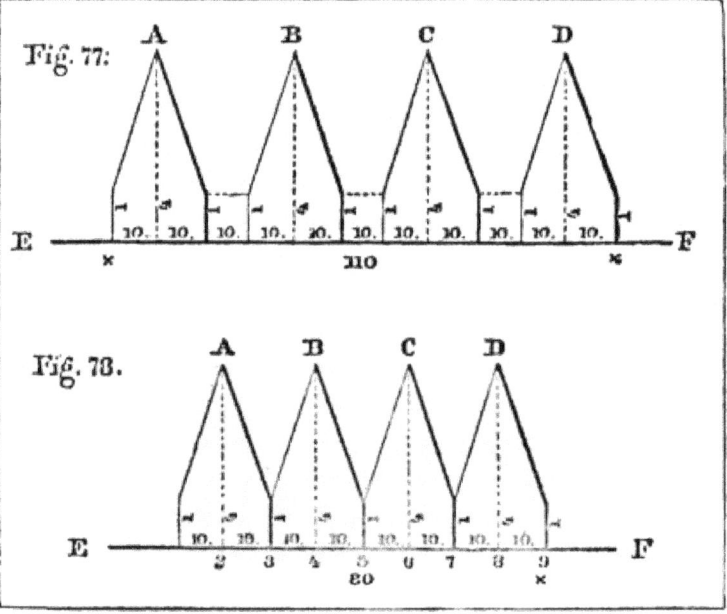

1. Suppose the solids to be separated from each other by the distance of 10 feet (or any other), and let each be computed independently by means of Simpson's Multipliers, or as they are all exactly alike, let one be computed and multiplied by 4, *as follows:*

	Cross-sec. in Sq. Ft.	Simpson's Mults.		Results in Sq. Ft.
This is clearly a *Prismoidal Computation.*	1	× 1	=	1
	4	× 4	=	16
	1	× 1	=	1
			6)18	
	Mean Area =		3 × 20	= 60 A.

$$60 \times 4 = 240 \text{ Cubic Feet} = A + B + C + D.$$

2. Now, suppose the solids to be slid along the base line EF, until they come in *actual contact* with each other, as shown in *Fig*. 78. Then it becomes evident that the intermediate sections at *odd* numbers (1, 3, etc.), which, in the detached solids, *Fig*. 77, were used but *once*, are here, when combined, to be used *twice;* while the mid-sections, or those at *even* numbers, are to be used *four times*, and the extreme end sections only *once* each; so that they become, in effect, when treated thus, *the Multipliers of Simpson;* while the divisor is changed to 3, because the common interval is reduced *one-half;*—and the volume of the four solids, when aggregated together, so as to form a single body, would be computed by Simpson's Rule, or by his Multipliers, *as follows:*

By Simpson's Rule, $\dfrac{2 + 64 + 6}{3} \times \overset{\text{Common Interval.}}{10} = 240$, *as above.*

	Secs.		Mults.		Sq. Ft.
	1	×	1	=	1
	4	×	4	=	16
	1	×	2	=	2
	4	×	4	=	16
By *Simpson's Multipliers*,	1	×	2	=	2
with 8 equal intervals.	4	×	4	=	16
	1	×	2	=	2
	4	×	4	=	16
	1	×	1	=	1
					3)72

General Mean Area . . = 24
Common Interval . . . = 10
Result same as before . . = 240 C. Feet.

As Simpson's Rule is an important one, we hope the above digression to explain it fully, and the foundation on which it rests, will be excused by the reader.

37. Having then taken off from a rough profile of the line run the centre hights to the nearest foot, and from the field notes ascertained the average surface slope at each station to the nearest 5°, we enter Tables 2, 3, and 4, and obtain the triangular areas to the intersection of the side-slopes (supposed to be prolonged to meet), to the nearest foot of area, for *rock cutting, earth cutting*, or *embankment*—each of these, that we may require, we set down separately in a column, and where a case occurs of a hight exceeding the limits of the Tables named, then we resort to the initial triangles of Table 1, by means of which the area due to any hight *whatever* may easily be ascertained; then, if we find we have an *even* number of equal stations, we apply Simpson's Multipliers to the column of areas, and speedily compute *the solidity.*

But if the equal intervals or stations are found to be *uneven* in number, strike off one station temporarily for independent calculation, and then the number of intervals becoming *even,* we are ready to apply Simpson's Multipliers, in a column parallel to that of areas, and beginning at 1, as 1, 4, 2, 4, 2, 4, etc., multiplying each cross-section by its proper factor, and placing the results in a third parallel column, which we sum up and divide the total by 3 (giving a Mean Area as the quotient), add to this the mean area of the station reserved (if any), which gives a General Mean Area, to be multiplied by the equal interval, or length of station—say 200 feet, or whatever distance has been adopted and used as a common interval or station —the result will be cubic feet, from which cubic yards (if desired) can easily be found.

But, inasmuch as the quotient of 3 (with the mean area of the reserved station (if any) added in) *is a General Mean Area*—usually in square feet—it will be found more convenient, and usually more accurate, to use it in connection with our Table 5, at the end of the Book, to find the cubic yards which may be desired, according to the directions preceding the Table.

We will now proceed to give examples of the process above explained, and for this purpose we will take *the adjacent bank and rock cut,* profiled on *Fig.* 76, *Art.* **24,** as being an appropriate example of this expeditious method of computing an embankment, or an excavation in a single body, with sufficient accuracy for the purpose contemplated, *and without unusual delay.*

Fig. 76. BANK.

Here we find the Bank to be 1000 feet in length between the grade points, or 5 intervals of 200 feet each; the number of intervals being *uneven,* we must temporarily omit one station to bring this case within the rule; let the station omitted, and to be calculated independently, be from 5 to 7 = 200 feet.

11

Tabulation.

Sta.	Areas.	Mults.	Sq. Feet.
1	24	× 1	= 24
3	495	× 4	= 1980
5 and 7 united.	3123	× 2	= 6246
9	1197	× 4	= 4788
11	24	× 1	= 24

$$3)\overline{13062}$$

4354   = Partial Mean Area.

Add area of reserved station.

The hight of the embankment and the surface-slope at 5 and 7 being the same, this reserved station is a *Prism*, of which the base, or sectional area, is 3123 square feet, and length = 200 feet . . . . = . 3123   = Mean Area, reserved
                                          station.

General Mean Area. . . =	7477	Square Feet.
	200	Common Interval.
Solidity . . . . =	1495400	Cubic Feet.
Or, ... =	55385	*Cubic Yards.*
Tabulated, by Roots and Squares, in 100 feet stations . =	55088	" "
Difference about the half of one per cent. more . . . . =	+297	" "

Tabulated by Roots and Squares in 100 feet stations, as though for a final estimate, the Bank in our example contains 55,088 Cubic Yards, while by our hasty process the result is 55,385 Cubic Yards, or 297 Cubic Yards *more*. As this difference is but little more *than the half of one per cent.* upon the true amount, it can hardly be considered as *excessive* for a method as brief and simple as that under consideration here.

Fig. 76. Rock-Cut.

The Rock-Cut, like the Bank connected with it, and tabulated above, is 1000 feet in length between the grade points, or 5 intervals of 200 feet each, which, being an *uneven* number, we must temporarily omit one station, and calculate it separately, to make the number of intervals *even,* and bring it within the scope of Simpson's Rule. Let the station reserved be from 19 to 21 = 200 feet.

Tabulation.

Sta.	Areas.	Mults.		Sq. Feet.
11	192	× 1	=	192
13	646	× 4	=	2584
15	975	× 2	=	1950
17	589	× 4	=	2356
19	771	× 1	=	771

$$3 \overline{)7853}$$

2618   = Partial Mean Area.

Station reserved from 19 to 21, to make the number of intervals *even,* as required by the Rule of Simpson.

$$\left\{ \begin{array}{l} 19 = 771 \times 1 = 771 \\ 20 = 433 \times 4 = 1732 \\ 21 = 192 \times 1 = 192 \\ \hline 6\overline{)2695} \\ \text{Mean Area} = 449 \end{array} \right\} = 449$$ Mean Area, reserved station.

General Mean Area . . . = 3067   Square Feet.

200   Common Interval.

Solidity . . . . = 613400 = 22718 Cubic Yards.

Tabulated by Roots and Squares, in stations of 100 feet = 623298 = 23085   "   "

Diff. about 1½ *per cent. less* = 9898 = —367   "   "

38. . . . . . It will be observed that in the preceding computations the *Grade Prism* is not taken into the account, as it is deductive on both sides, and the only object in hand *is a comparison.*

The triangular section, or area of the Grade Prism, *is the minimum area found,* in the methods of computation which go down to the junction of the side-slopes, and always occurs when the road-bed comes to grade, or the level hight on the centre line is 0.

And we *repeat,* it is necessary to be careful that the volume of the Grade Prism (always included in the earlier steps of such calculations) is duly deducted before the close of the operation, in order to determine *the solidity above* the road-bed in cutting, or *below* it in filling.

We may here add that the earth cutting profiled *ante*, and there correctly computed by Roots and Squares, if calculated with Simpson's Multipliers by the hasty process above given, in stations of 200 feet, as though it were part of an *exploring line*, would give *as follows :*

Volume of Grade Prism omitted in both.

C. Yards.

$$\left\{ \begin{array}{l} \textit{Tabulated ante}, \text{ in 100 feet stations} \ . \ . \ . \ . \ . \ . \ . \ = 18684 \\ \quad\quad\text{``}\quad\quad \text{by our Hasty Process, 200 feet stations} \ . \ . \ . = 18378 \\ \textit{Difference} \text{ about } 1\tfrac{1}{2} \text{ per cent. } \textit{less} \ . \ . \ . \ . \ . \ . \ . \ . = \overline{\quad 306} \end{array} \right.$$

So that this brief and hasty process, being *very expeditious and proximately correct* (usually varying only 1 or 2 per cent. from the truth), may be safely accepted as adequate for the determination of the quantities of earthwork, which may be needed *in rough estimates, or for the comparison of exploring lines.*

For the purpose of furnishing additional aid in expediting Preliminary Estimates, we annex four small Tables, which will be found quite convenient.

TABLES

1, 2, 3, and 4.

For use in Hasty or Preliminary Estimates.

Viz: 1. Initial Triangles to a hight of *unity*, and various side and surface slopes.

Triangular Areas to Intersection of Slopes.

	Side-slopes.			Surface-slopes.		
2. Rock Cut ½ to 1, and	0°,	5°,	10°,	15°,	20°.	
3. Earth Cut 1 to 1, and	"	"	"	"	"	
4. Embankment and	"	"	"	"	"	

In using Tables 2, 3, and 4, the centre hight is generally to be taken to the nearest foot (though tenths might be used), and the ground surface slope to the nearest 5°—these being thought sufficient for rough estimates—and if the centre hight should exceed the limits of the Tables, then, by using the Initial Triangles of Table 1, the area of the cross-section for any hight *whatever* can be easily ascertained. If the centre hights necessarily contain tenths of feet, they may be proportioned for by the columns in the Tables for that purpose.

Note.—All the triangular areas in Tables 2, 3, and 4, extend from ground line to junction of side-slopes *prolonged*, or edge of the diedral angle, which, with ground surface, bounds on every side the earthwork solid. The road-bed, or grade line, may be assumed to cross the triangle at any given distance from the angle of intersection; but the volume of the Grade Prism must always be ascertained and deducted at the close of the operation, in every calculation involving the triangular areas of the Tables. The altitude of the Grade Triangle is invariably = road-bed ÷ 2 r, and its area will be found opposite to this hight in the 0 column of the Tables.

TABLE 1.

Initial Triangles, to a hight of unity, with side-slopes of ½ to 1 for Rock; 1 to 1 for earth; 1½ to 1 for embankment; and ground surface slopes of 0°, 5°, 10°, 15°, 20°. All computed to six places of decimals, and all extending from ground line to intersection of side-slopes.

Side-slopes.				Ground Surface-slopes.				
		Cot.	Tan.	0°	5°	10°	15°	20°
Ratio.	Angle.	of Trian. Tables.		Tan. = ·0	Tan. = ·0875	Tan. = ·1763	Tan. = ·2679	Tan. = ·3640
½ to 1	71° 34′	0·3333	3	0·333333	0·333586	0·334457	0·335682	0·338289
1 to 1	45°	1	1	1	1·007713	1·032088	1·077359	1·152663
1½ to 1	33° 41′	1·5	·6666	1·5	1·526688	1·613298	1·790002	2·137798

Note.—A similar Table may easily be extended to any other side, or surface-slope, and such extension would often be found useful to the engineer.

Application of the above Table.

Rule.—For any given hight, to find the triangular area, when conditioned as above.

Multiply the Square of the Given Hight by the Tabular Area of the Initial Triangle.

Example.

Let the given hight be 26·4 feet, the side-slope 1 to 1, and the ground surface-slope 20°.

Then, $(26·4)^2 \times 1·152663 = 803·36$ square feet = area of triangle required.

Triangular Areas, in square feet, for side-slopes of ⅓ to 1, *to intersection of slopes.* (*r* = ⅓.) Slope angle = 71° 34′.

TABLE 2—Rock-cut.

Hight in feet	Surf.-slope 0°.		Surf.-slope 5°.		Surf.-slope 10°.		Surf.-slope 15°.		Surf.-slope 20°.		Hight in feet
	Areas.	Pro. for ·1.	Areas.	Pro. for ·1.	Areas.	Pro. for ·1.	Areas.	Pro. for ·1.	Areas.	Pro. for ·1.	
1	·3333	·03	·3336	·03	·3345	·03	·3357	·03	·3385	·03	1
2	1·3333	·10	1·3	·10	1·3	·10	1·3	·10	1·4	·10	2
3	3	·17	3	·17	3	·17	3	·17	3	·17	3
4	5·3333	·23	5	·23	5	·23	5	·23	5	·24	4
5	8·3333	·30	8	·30	8	·30	8	·30	8	·30	5
6	12	·37	12	·37	12	·37	12	·37	12	·37	6
7	16·3333	·43	16	·43	16	·43	16	·44	17	·44	7
8	21·3333	·50	21	·50	21	·50	22	·50	22	·51	8
9	27	·57	27	·57	27	·57	27	·57	28	·58	9
10	33·3333	·63	34	·63	33	·64	34	·64	34	·64	10
11	40·3333	·70	40	·70	41	·70	41	·71	41	·71	11
12	48	·77	48	·77	48	·77	48	·77	49	·78	12
13	56·3333	·83	56	·83	57	·84	57	·84	57	·88	13
14	65·3333	·90	65	·90	66	·90	66	·91	66	·91	14
15	75	·97	75	·97	75	·97	76	·98	76	·98	15
16	85·3333	1·03	85	1·03	86	1·04	86	1·04	87	1·05	16
17	96·3333	1·10	96	1·10	97	1·10	97	1·11	98	1·11	17
18	108	1·17	108	1·17	108	1·17	109	1·18	110	1·18	18
19	120·3333	1·23	121	1·23	121	1·24	121	1·24	122	1·25	19
20	133·3333	1·30	133	1·30	134	1·30	135	1·30	135	1·31	20
21	147	1·37	147	1·37	148	1·37	148	1·37	149	1·38	21
22	161·3333	1·43	161	1·43	162	1·44	163	1·44	164	1·45	22
23	176·3333	1·50	176	1·50	176	1·50	179	1·51	179	1·52	23
24	192	1·57	192	1·57	193	1·57	194	1·58	195	1·59	24
25	208·3333	1·63	209	1·63	202	1·64	210	1·64	212	1·66	25
26	225·3333	1·70	226	1·70	226	1·70	227	1·71	229	1·72	26
27	243	1·77	244	1·77	244	1·77	245	1·78	247	1·79	27
28	261·3333	1·83	262	1·84	262	1·84	263	1·85	265	1·85	28
29	280·3333	1·90	281	1·90	281	1·91	282	1·91	285	1·93	29
30	300	1·97	300	1·97	301	1·97	302	1·98	305	2·00	30
31	320·3333	2·03	321	2·04	322	2·04	323	2·05	325	2·06	31
32	341·3333	2·10	342	2·10	343	2·11	344	2·12	346	2·13	32
33	364	2·17	363	2·17	366	2·17	366	2·18	368	2·20	33
34	385·3333	2·23	386	2·24	387	2·24	388	2·25	391	2·27	34
35	408·3333	2·30	409	2·30	410	2·31	412	2·32	415	2·34	35
36	432	2·37	433	2·37	434	2·38	436	2·39	439	2·40	36
37	456·3333	2·43	457	2·44	458	2·44	460	2·45	463	2·47	37
38	481·3333	2·50	482	2·50	483	2·51	485	2·52	489	2·54	38
39	507	2·57	508	2·57	509	2·58	511	2·59	515	2·61	39
40	533·3333	2·63	534	2·64	535	2·64	538	2·66	541	2·67	40
41	560·3333	2·70	561	2·70	562	2·71	565	2·72	569	2·74	41
42	588	2·77	589	2·77	590	2·77	593	2·79	597	2·81	42
43	616·3333	2·83	617	2·84	618	2·84	621	2·86	625	2·88	43
44	645·3333	2·90	646	2·90	648	2·91	651	2·92	655	2·94	44
45	675	2·97	676	2·97	677	2·98	680	2·99	685	3·01	45
46	705·3333	3·03	706	3·04	708	3·04	711	3·06	716	3·08	46
47	736·3333	3·10	737	3·10	742	3·11	742	3·13	747	3·15	47
48	768	3·17	769	3·17	771	3·18	774	3·19	780	3·21	48
49	800·3333	3·23	801	3·24	805	3·24	807	3·26	812	3·28	49
50	833·3333	3·30	834	3·31	836	3·31	840	3·33	846	3·35	50
Hight in feet	Surf.-slope 0°.		Surf.-slope 5°.		Surf.-slope 10°.		Surf.-slope 15°.		Surf.-slope 20°.		Hight in feet

Triangular Areas, in square feet, for side-slopes of 1 to 1, *to intersection of slopes.* (r = 1.) Slope angle = ·45°.

TABLE 3—Earth-cut.

Hight in feet	Surf.-slope 0°		Surf.-slope 5°		Surf.-slope 10°		Surf.-slope 15°		Surf.-slope 20°		Hight in feet
	Areas.	Pro. for ·1.	Areas.	Pro. for ·1.	Areas.	Pro. for ·1.	Areas.	Pro. for ·1.	Areas.	Pro. for ·1.	
1	1·000	·10	1·0077	·10	1·0321	·10	1·0773	·11	1.1527	·12	1
2	4	·30	4	·30	4	·31	4	·32	5	·35	2
3	9	·50	9	·50	9	·52	10	·54	11	·58	3
4	16	·70	16	·70	17	·72	17	·75	18	·81	4
5	25	·90	25	·90	26	·93	27	·97	29	1·04	5
6	36	1·10	36	1·11	37	1·14	39	1·19	42	1·27	6
7	49	1·30	49	1·31	51	1·34	53	1·40	56	1·50	7
8	64	1·50	64	1·51	66	1·55	69	1·62	74	1·73	8
9	81	1·70	82	1·71	84	1·75	87	1·83	93	1·96	9
10	100	1·90	101	1·91	103	1·96	108	2·05	115	2·19	10
11	121	2·10	122	2·12	125	2·17	130	2·26	139	2·42	11
12	144	2·30	145	2·32	149	2·37	155	2·48	166	2·65	12
13	169	2·50	170	2·52	174	2·58	182	2·69	195	2·88	13
14	196	2·70	198	2·72	202	2·79	211	2·91	226	3·11	14
15	225	2·90	227	2·92	232	2·99	242	3·12	259	3·34	15
16	256	3·10	258	3·12	264	3·20	276	3·34	295	3·57	16
17	289	3·30	291	3·33	298	3·41	311	3·56	333	3·80	17
18	324	3·50	327	3·53	334	3·61	349	3·77	373	4·03	18
19	361	3·70	364	3·73	373	3·82	389	3·99	416	4·27	19
20	400	3·90	403	3·93	413	4·02	431	4·20	461	4·50	20
21	441	4·10	444	4·13	455	4·23	475	4·42	508	4·73	21
22	484	4·30	488	4·33	499	4·44	521	4·63	558	4·96	22
23	529	4·50	533	4·53	546	4·64	570	4·85	610	5·19	23
24	576	4·70	580	4·74	594	4·85	621	5·06	664	5·42	24
25	625	4·90	630	4·94	645	5·06	673	5·28	720	5·65	25
26	676	5·10	681	5·14	698	5·26	728	5·49	779	5·88	26
27	729	5·30	735	5·34	752	5·47	785	5·71	840	6·11	27
28	784	5·50	790	5·54	809	5·68	845	5·92	904	6·34	28
29	841	5·70	848	5·74	868	5·88	906	6·14	969	6·57	29
30	900	5·90	907	5·95	929	6·09	970	6·36	1037	6·80	30
31	961	6·10	968	6·15	992	6·30	1035	6·57	1108	7·03	31
32	1024	6·30	1032	6·35	1057	6·50	1103	6·79	1180	7·26	32
33	1089	6·50	1097	6·55	1124	6·71	1173	7·00	1255	7·49	33
34	1156	6·70	1165	6·75	1193	6·91	1245	7·22	1333	7·72	34
35	1225	6·90	1234	6·95	1264	7·12	1320	7·43	1412	7·95	35
36	1296	7·10	1306	7·15	1338	7·33	1396	7·65	1494	8·18	36
37	1369	7·30	1380	7·36	1413	7·53	1475	7·86	1578	8·41	37
38	1444	7·50	1455	7·56	1490	7·74	1556	8·08	1665	8·64	38
39	1521	7·70	1533	7·76	1570	7·95	1639	8·29	1753	8·88	39
40	1600	7·90	1612	7·96	1651	8·15	1724	8·51	1844	9·11	40
41	1681	8·10	1694	8·16	1735	8·36	1811	8·73	1938	9 34	41
42	1764	8·30	1778	8·36	1820	8·57	1900	8 94	2033	9·57	42
43	1849	8·50	1863	8·56	1908	8 77	1992	9·16	2131	9·80	43
44	1936	8·70	1951	8·77	1998	8·98	2086	9·37	2232	10·03	44
45	2025	8·90	2041	8·97	2090	9·18	2182	9·59	2334	10·26	45
46	2116	9·10	2132	9·17	2184	9·39	2280	9·80	2439	10·49	46
47	2209	9·30	2226	9·37	2280	9·60	2380	10·02	2546	10·72	47
48	2304	9·50	2322	9·57	2378	9·80	2482	10·23	2656	10·95	48
49	2401	9·70	2420	9·77	2478	10·01	2587	10·45	2768	11·18	49
50	2500	9·90	2519	9·97	2580	10·22	2693	10·67	2882	11·41	50
Hight in feet	Surf.-slope 0°		Surf.-slope 5°		Surf.-slope 10°		Surf.-slope 15°		Surf.-slope 20°		Hight in feet

Triangular areas, in square feet, for side-slopes of 1½ to 1, *to intersection of slopes.* (*r* = 1½.) Slope angle = 33° 41′.

TABLE 4—Bank.

Hight in feet.	Surf.-slope 0°.		Surf.-slope 5°.		Surf.-slope 10°.		Surf.-slope 15°.		Surf.-slope 20°.		Hight in feet.
	Areas.	Pro. for 1.	Areas.	Pro. for 1.	Areas.	Pro. for 1.	Areas.	Pro. for 1.	Areas.	Pro. for 1.	
1	1·5000	·15	1·5267	·15	1·6133	·16	1·7900	·18	2·1378	·24	1
2	6	·45	6	·46	6	·48	7	·54	9	·64	2
3	13·5	·75	14	·76	15	·81	16	·89	19	1·07	3
4	24	1·05	25	1·07	26	1·13	29	1·25	34	1·50	4
5	37·5	1·35	38	1·37	40	1·45	45	1·61	54	1·92	5
6	54	1·65	55	1·68	58	1·78	64	1·97	77	2·35	6
7	73·5	1·95	75	1·98	79	2·10	88	2·33	105	2·78	7
8	96	2·25	98	2·29	103	2·42	115	2·68	137	3·21	8
9	121·5	2·55	124	2·59	131	2·74	145	3·04	173	3·63	9
10	150	2·85	153	2·90	161	3·06	179	3·39	214	4·06	10
11	181·5	3·15	185	3·20	195	3·39	217	3·74	259	4·49	11
12	216	3·45	220	3·51	232	3·71	258	4·12	308	4·92	12
13	253·5	3·75	258	3·82	273	4·03	302	4·47	361	5·34	13
14	294	4·05	299	4·12	316	4·36	351	4·83	419	5·77	14
15	337·5	4·35	344	4·43	363	4·68	403	5·19	481	6·20	15
16	384	4·65	391	4·73	413	5·00	458	5·55	547	6·63	16
17	433·5	4·95	441	5·04	466	5·32	517	5·92	618	7·05	17
18	486	5·25	495	5·34	523	5·65	580	6·26	693	7·48	18
19	541·5	5·55	551	5·65	582	5·97	646	6·62	772	7·91	19
20	600	5·85	611	5·96	645	6·29	716	6·98	855	8·34	20
21	661·5	6·15	673	6·26	711	6·61	789	7·34	943	8·76	21
22	726	6·45	739	6·56	781	6·94	866	7·69	1035	9·19	22
23	793·5	6·75	808	6·87	853	7·26	947	8·05	1131	9·62	23
24	864	7·05	879	7·17	929	7·58	1031	8·41	1231	10·05	24
25	937·5	7·35	954	7·48	1008	7·90	1118	8·77	1336	10·47	25
26	1014	7·65	1032	7·79	1090	8·23	1210	9·13	1445	10·90	26
27	1093·5	7·95	1113	8·09	1176	8·55	1304	9·48	1558	11·33	27
28	1176	8·25	1197	8·40	1265	8·87	1403	9·84	1676	11·76	28
29	1261·5	8·55	1284	8·70	1357	9·19	1505	10·20	1798	12·18	29
30	1350	8·85	1374	9·00	1452	9·52	1610	10·55	1924	12·61	30
31	1441·5	9·15	1467	9·31	1550	9·84	1719	10·91	2054	13·04	31
32	1536	9·45	1563	9·62	1652	10·16	1832	11·27	2189	13·47	32
33	1633·5	9·75	1662	9·92	1757	10·48	1948	11·63	2328	13·89	33
34	1734	10·05	1765	10·23	1865	10·81	2068	11·99	2471	14·32	34
35	1837·5	10·35	1870	10·53	1976	11·13	2192	12·35	2619	14·75	35
36	1944	10·65	1978	10·84	2090	11·45	2319	12·70	2770	15·18	36
37	2053·5	10·95	2090	11·14	2208	11·77	2449	13·06	2926	15·60	37
38	2166	11·25	2204	11·45	2329	12·10	2584	13·42	3087	16·03	38
39	2281·5	11·55	2322	11·76	2453	12·42	2721	13·78	3251	16·46	39
40	2400	11·85	2442	12·06	2581	12·74	2863	14·14	3420	16·89	40
41	2521·5	12·15	2566	12·36	2711	13·06	3008	14·50	3593	17·31	41
42	2646	12·45	2693	12·67	2845	13·39	3156	14·85	3771	17·74	42
43	2773·5	12·75	2823	12·98	2982	13·71	3308	15·21	3952	18·17	43
44	2904	13·05	2955	13·28	3123	14·03	3464	15·57	4138	18·60	44
45	3037·5	13·35	3091	13·59	3266	14·35	3623	15·92	4329	19·02	45
46	3174	13·65	3230	13·89	3413	14·68	3786	16·28	4523	19·45	46
47	3313·5	13·95	3372	14·20	3563	15·00	3952	16·64	4722	19·88	47
48	3456	14·25	3517	14·50	3716	15·32	4122	16·99	4925	20·31	48
49	3601·5	14·55	3665	14·81	3873	15·64	4296	17·35	5132	20·74	49
50	3750	14·85	3816	15·12	4032	15·97	4473	17·71	5344	21·16	50
Hight in feet.	Surf.-slope 0°.		Surf.-slope 5°.		Surf.-slope 10°.		Surf.-slope 15°.		Surf.-slope 20°.		Hight in feet.

TABLE OF CUBIC YARDS

IN FULL STATIONS, OR LENGTHS OF 100 FEET.

CALCULATED FOR EVERY FOOT AND TENTH OF MEAN AREA,

FROM 0· TO 1000· SUPERFICIAL FEET.

Note.—On every page of the Table, the columns on both sides headed M.A. contain the Mean Areas, in square, or superficial feet.

The horizontal lines at top and bottom show the tenths of square feet of Mean Area.

And the figures in the body of the Table, computed to three places of decimals, are the Cubic Yards (for 100· feet), corresponding to the feet and tenths of Mean Area, indicated in the side columns, and lines of tenths at top and bottom.

EXPLANATION OF THE TABLE OF CUBIC YARDS,

To Mean Areas, in lengths of 100· feet, and of its Applications.

This Table is computed to facilitate the conversion into *Cubic Yards* of the content of any solid 100 feet in length, of which the *Mean Area* in superficial feet has been ascertained. It applies *directly* to all Mean Areas from 0· to 1000· square feet (including tenths of feet), and being calculated to three decimal places, it extends *indirectly* to 100,000· superficial feet of Mean Area, as will be shown hereafter.

EXAMPLE 1.
———
Cubic yards for *full stations* (100·)

To find the Cubic Yards, belonging to 579·⁸ sup. ft. of Mean Area, *for a full station*, or length of 100· feet:

Opposite 579· and under ·8 we find the content, or *solidity*...............=2147·407 cubic yards. Which is equal to

579·⁸ sq. ft. of Mean Area × 100· feet long, and divided by 27.

EXAMPLE 2.
—
Cubic yards for
short stations
(− 100·)

> Let the Mean Area of any solid, be 98·7 sq. ft. and its length 84 ft. lineal: (*being a short station*). Then at 98·7 we find 365·556 cubic yards, which being multiplied by ·84 taken decimally, gives 365·556 × ·84.........=307·067 cubic yards.
> Equal to... $\dfrac{98\cdot7 \times 84}{27\cdot}$.

EXAMPLE 3.
—
Cubic yards for
long stations
(+ 100·)

> Again, let the Mean Area be 88·6 and the length 259· feet (*or a long station*); then for 88·6 sq. ft. of Mean Area, we have 328·148 cubic yards, which multiplied by 2·59 (decimal) gives.........................=849·903 cubic yards.
> Equal to... $\dfrac{88\cdot6 \times 259\cdot}{27\cdot}$

This Table is especially useful in the computation of the Earthwork of Railroads, and other Public Works, where cross-sections have been taken normal to a guide line, at distances (generally) of 100· lineal feet (or full stations), and the Mean Area calculated in superficial feet and parts: but it is also applicable to any solid of which the mean section is known in square feet, and the length 100· feet, or any decimal part thereof.

For, if the distances apart of cross-sections, or lengths of stations, be more, or less, than 100· feet, we have only to take them *decimally*, as in the above examples, and by a simple multiplication, of the tabular quantity, belonging to the known area, the correct number of cubic yards will be ascertained.

The Table being calculated to *three* places of decimals, readily admits of being used for Mean Areas, much exceeding its direct range of 1000· superficial feet (as follows):

EXAMPLE 4. Suppose the Mean Area to be 98,967·4 sq. ft. (representing a cut 98·9 feet deep, and 1000· feet wide).

1 ....

> Then for 98,900· (by moving the decimal point of the tabular quantity of cubic yards for 989· two figures to the right)—
> We have, area 98,900· = 366,296·3 cubic yds.
> Add........ 67·4 = 249·6 " "
> Total, for sq. ft... 98,967·4 = 366,545·9 " "
> Equal to... $\dfrac{98,967\cdot4 \times 100}{27}$.

Again, take a Mean Area, of 100,048·⁹ sq. ft. (representing a cut 100· feet deep, and 1000· feet wide).

$$
\textit{2} \ldots\ldots\ldots\ldots \begin{cases}
\text{Then for 100,000 sq. ft. (by moving the deci-} \\
\text{mal point of the tabular quantity of cubic yards} \\
\text{for 1000· two figures to the right),} \\
\text{We have, 100,000 Area} = 370,370·⁴ \text{ cub. yds.} \\
\underline{\text{Add} \qquad 48·⁹ \text{ `` } = \qquad 181·¹ \text{ `` } \text{ ``}} \\
\text{Total for.....100,048·⁹ `` } = \overline{370,551·⁵} \text{ `` } \text{ ``} \\
\text{Equal to...} \dfrac{100,048·⁹ \times 100}{27}.
\end{cases}
$$

Example 4, shows the easy application of the Table, to Mean Areas, which may be called *immense*, by merely moving the decimal point, and a simple addition, as shown above.

Other methods of using the Table will occur to the reader, but the examples given seem sufficient for illustration.

Much pains have been taken to make this Table correct, to the nearest decimal, and we believe it may be safely depended on.

Note.—Besides its special application to Earthworks, the extensive Table following is also a general Table for the conversion of *any sum of Cubic Feet* into Cubic Yards. Thus, in the example at page 103, the reduced quantities of Cubic Feet sum up 227,200 — 30,000 = 197,200 Cubic Feet.

In such cases we have only to cut off two figures from the right (or ÷ by 100), and we have 1972, *the mean area,* which, in 100 feet length, would have produced the quantity given.

With 197·2 we enter the Table following, and find 730·370 Cubic Yards; now, moving the decimal point one place to the right, we have 7303·70 Cubic Yards, or in round numbers, 7304 Cubic Yards, as already given on page 103.

In like manner the Cubic Yards for *any* sum whatever of Cubic Feet can readily be obtained, and the Table being in itself strictly correct, *the result will be reliable.*

TABLE OF CUBIC YARDS, in full Stations, or lengths of 100 feet; for every foot and tenth of Mean Area, from 0 to 1000 Superficial Feet.

M.A.	·0	·1	·2	·3	·4	·5	·6	·7	·8	·9	M.A.
0	0·000	0·370	0·741	1·111	1·481	1·852	2·222	2·593	2·963	3·333	0
1	3·704	4·074	4·444	4·815	5·185	5·556	5·926	6·296	6·667	7·037	1
2	7·407	7·778	8·148	8·519	8·889	9·259	9·630	10·	10·370	10·741	2
3	11·111	11·481	11·852	12·222	12·593	12·963	13·333	13·704	14·074	14·444	3
4	14·815	15·185	15·556	15·926	16·296	16·667	17·037	17·407	17·778	18·148	4
5	18·519	18·889	19·259	19·630	20·	20·370	20·741	21·111	21·481	21·852	5
6	22·222	22·593	22·963	23·333	23·704	24·074	24·444	24·815	25·185	25·556	6
7	25·926	26·296	26·667	27·037	27·407	27·778	28·148	28·519	28·889	29·259	7
8	29·630	30·	30·370	30·741	31·111	31·481	31·852	32·222	32·593	32·963	8
9	33·333	33·704	34·074	34·444	34·815	35·185	35·556	35·926	36·296	36·667	9
10	37·037	37·407	37·778	38·148	38·519	38·889	39·259	39·630	40·	40·370	10
11	40·741	41·111	41·481	41·852	42·222	42·593	42·963	43·333	43·704	44·074	11
12	44·444	44·815	45·185	45·556	45·926	46·296	46·667	47·037	47·407	47·778	12
13	48·148	48·519	48·889	49·259	49·630	50·	50·370	50·741	51·111	51·481	13
14	51·852	52·222	52·593	52·963	53·333	53·704	54·074	54·444	54·815	55·185	14
15	55·556	55·926	56·296	56·667	57·037	57·407	57·778	58·148	58·519	58·889	15
16	59·259	59·630	60·	60·370	60·741	61·111	61·481	61·852	62·222	62·593	16
17	62·963	63·333	63·704	64·074	64·444	64·815	65·185	65·556	65·926	66·296	17
18	66·667	67·037	67·407	67·778	68·148	68·519	68·889	69·259	69·630	70·	18
19	70·370	70·741	71·111	71·481	71·852	72·222	72·593	72·963	73·333	73·704	19
20	74·074	74·444	74·815	75·185	75·556	75·926	76·296	76·667	77·037	77·407	20
21	77·778	78·148	78·519	78·889	79·259	79·630	80·	80·370	80·741	81·111	21
22	81·481	81·852	82·222	82·593	82·963	83·333	83·704	84·074	84·444	84·815	22
23	85·185	85·556	85·926	86·296	86·667	87·037	87·407	87·778	88·148	88·519	23
24	88·889	89·259	89·630	9 0·	90·370	90·741	91·111	91·481	91·852	92·222	24
25	92·593	92·963	93·333	93·704	94·074	94·444	94·815	95·185	95·556	95·926	25
26	96·296	96·667	97·037	97·407	97·778	98·148	98·519	98·889	99·259	99·630	26
27	100·	100·370	100·741	101·111	101·481	101·852	102·222	102·593	102·963	103·333	27
28	103·704	104·074	104·444	104·815	105·185	105·556	105·926	106·296	106·667	107·037	28
29	107·407	107·778	108·148	108·519	108·889	109·259	109·630	110·	110·370	110·741	29
30	111·111	111·481	111·852	112·222	112·593	112·963	113·333	113·704	114·074	114·444	30
31	114·815	115·185	115·556	115·926	116·296	116·667	117·037	117·407	117·778	118·148	31
32	118·519	118·889	119·259	119·630	120·	120·370	120·741	121·111	121·481	121·852	32
33	122·222	122·593	122·963	123·333	123·704	124·074	124·444	124·815	125·185	125·556	33
34	125·926	126·296	126·667	127·037	127·407	127·778	128·148	128·519	128·889	129·259	34
35	129·630	130·	130·370	130·741	131·111	131·481	131·852	132·222	132·593	132·963	35
36	133·333	133·704	134·074	134·444	134·815	135·185	135·556	135·926	136·296	136·667	36
37	137·037	137·407	137·778	138·148	138·519	138·889	139·259	139·630	140·	140·370	37
38	140·741	141·111	141·481	141·852	142·222	142·593	142·963	143·333	143·704	144·074	38
39	144·444	144·815	145·185	145·556	145·926	146·296	146·667	147·037	147·407	147·778	39
40	148·148	148·519	148·889	149·259	149·630	150·	150·370	150·741	151·111	151·481	40
41	151·852	152·222	152·593	152·963	153·333	153·704	154·074	154·444	154·815	155·185	41
42	155·556	155·926	156·296	156·667	157·037	157·407	157·778	158·148	158·519	158·889	42
43	159·259	159·630	160·	160·370	160·741	161·111	161·481	161·852	162·222	162·593	43
44	162·963	163·333	163·704	164·074	164·444	164·815	165·185	165·556	165·926	166·296	44
45	166·667	167·037	167·407	167·778	168·148	168·519	168·889	169·259	169·630	170·	45
46	170·370	170·741	171·111	171·481	171·852	172·222	172·593	172·963	173·333	173·704	46
47	174·074	174·444	174·815	175·185	175·556	175·926	176·296	176·667	177·037	177·407	47
48	177·778	178·148	178·519	178·889	179·259	179·630	180·	180·370	180·741	181·111	48
49	181·481	181·852	182·222	182·593	182·963	183·333	183·704	184·074	184·444	184·815	49
50	185·185	185·556	185·926	186·296	186·667	187·037	187·407	187·778	188·148	188·519	50
51	188·889	189·259	189·630	190·	190·370	190·741	191·111	191·481	191·852	192·222	51
52	192·593	192·963	193·333	193·704	194·074	194·444	194·815	195·185	195·556	195·926	52
53	196·296	196·667	197·037	197·407	197·778	198·148	198·519	198·889	199·259	199·630	53
54	200·	200·370	200·741	201·111	201·481	201·852	202·222	202·593	202·963	203·333	54
55	203·704	204·074	204·444	204·815	205·185	205·556	205·926	206·296	206·667	207·037	55
56	207·407	207·778	208·148	208·519	208·889	209·259	209·630	210·	210·370	210·741	56
57	211·111	211·481	211·852	212·222	212·593	212·963	213·333	213·704	214·074	214·444	57
58	214·815	215·185	215·556	215·926	216·296	216·667	217·037	217·407	217·778	218·148	58
59	218·519	218·889	219·259	219·630	220·	220·370	220·741	221·111	221·481	221·852	59
60	222·222	222·593	222·963	223·333	223·704	224·074	224·444	224·815	225·185	225·556	60
M.A.	·0	·1	·2	·3	·4	·5	·6	·7	·8	·9	M.A.

MEAN AREAS 0 to 60.

CUBIC YARDS TO MEAN AREAS FOR 100 FEET IN LENGTH.

M.A.	·0	·1	·2	·3	·4	·5	·6	·7	·8	·9	M.A.
61	225·926	226·296	226·667	227·037	227·407	227·778	228·148	228·519	228·889	229·259	61
62	229·630	250·	230·370	230·741	231·111	231·481	231·852	232·222	232·593	232·963	62
63	233·333	233·704	234·074	234·444	234·815	235·185	235·556	235·926	236·296	236·067	63
64	237·037	237·407	237·778	238·148	238·519	238·889	239·259	239·630	240·	240·370	64
65	240·741	241·111	241·481	241·852	242·222	242·593	242·963	243·333	243·704	244·074	65
66	244·444	244·815	245·185	245·556	245·926	246·296	246·667	247·037	247·407	247·778	66
67	248·148	248·519	248·889	249·259	249·630	250·	250·370	250·741	251·111	251·481	67
68	251·852	252·222	252·593	252·963	253·333	253·704	254·074	254·444	254·815	255·185	68
69	255·556	255·926	256·296	256·667	257·037	257·407	257·778	258·148	258·519	258·889	69
70	259·259	259·630	260·	260·370	260·741	261·111	261·481	261·852	262·222	262·593	70
71	262·963	263·333	263·704	264·074	264·444	264·815	265·185	265·556	265·926	266·296	71
72	266·667	267·037	267·407	267·778	268·148	268·519	268·889	269·259	269·630	270·	72
73	270·370	270·741	271·111	271·481	271·852	272·222	272·593	272·963	273·333	273·704	73
74	274·074	274·444	274·815	275·185	275·556	275·926	276·296	276·667	277·037	277·407	74
75	277·778	278·148	278·519	278·889	279·259	279·630	280·	280·370	280·741	281·111	75
76	281·481	281·852	282·222	282·593	282·963	283·333	283·704	284·074	284·444	284·815	76
77	285·185	285·556	285·926	286·296	286·667	287·037	287·407	287·778	288·148	288·519	77
78	288·889	289·259	289·630	290·	290·370	290·741	291·111	291·481	291·852	292·222	78
79	292·593	292·963	293·333	293·704	294·074	294·444	294·815	295·185	295·556	295·926	79
80	296·296	296·667	297·037	297·407	297·778	298·148	298·519	298·889	299·259	299·630	80
81	300·	300·370	300·741	301·111	301·481	301·852	302·222	302·593	302·963	303·333	81
82	303·704	304·074	304·444	304·815	305·185	305·556	305·926	306·296	306·667	307·037	82
83	307·407	307·778	308·148	308·519	308·889	309·259	309·630	310·	310·370	310·741	83
84	311·111	311·481	311·852	312·222	312·593	312·963	313·333	313·704	314·074	314·444	84
85	314·815	315·185	315·556	315·926	316·296	316·667	317·037	317·407	317·778	318·148	85
86	318·519	318·889	319·259	319·630	320·	320·370	320·741	321·111	321·481	321·852	86
87	322·222	322·593	322·963	323·333	323·704	324·074	324·444	324·815	325·185	325·556	87
88	325·926	326·296	326·667	327·037	327·407	327·778	328·148	328·519	328·889	329·259	88
89	329·630	330·	330·370	330·741	331·111	331·481	331·852	332·222	332·593	332·963	89
90	333·333	333·704	334·074	334·444	334·815	335·185	335·556	335·926	336·296	336·667	90
91	337·037	337·407	337·778	338·148	338·519	338·889	339·259	339·630	340·	340·370	91
92	340·741	341·111	341·481	341·852	342·222	342·593	342·963	343·333	343·704	344·074	92
93	344·444	344·815	345·185	345·556	345·926	346·296	346·667	347·037	347·407	347·778	93
94	348·148	348·519	348·889	349·259	349·630	350·	350·370	350·741	351·111	351·481	94
95	351·852	352·222	352·593	352·963	353·333	353·704	354·074	354·444	354·815	355·185	95
96	355·556	355·926	356·296	356·667	357·037	357·407	357·778	358·148	358·519	358·889	96
97	359·259	359·630	360·	360·370	360·741	361·111	361·481	361·852	362·222	362·593	97
98	362·963	363·333	363·704	364·074	364·444	364·815	365·185	365·556	365·926	366·296	98
99	366·667	367·037	367·407	367·778	368·148	368·519	368·889	369·259	369·630	370·	99
100	370·370	370·741	371·111	371·481	371·852	372·222	372·593	372·963	373·333	373·704	100
101	374·074	374·444	374·815	375·185	375·556	375·926	376·296	376·667	377·037	377·407	101
102	377·778	378·148	378·519	378·889	379·259	379·630	380·	380·370	380·741	381·111	102
103	381·481	381·852	382·222	382·593	382·963	383·333	383·704	384·074	384·444	384·815	103
104	385·185	385·556	385·926	386·296	386·667	387·037	387·407	387·778	388·148	388·519	104
105	388·889	389·259	389·630	390·	390·370	390·741	391·111	391·481	391·852	392·222	105
106	392·593	392·963	393·333	393·704	394·074	394·444	394·815	395·185	395·556	395·926	106
107	396·296	396·667	397·037	397·407	397·778	398·148	398·519	398·889	399·259	399·630	107
108	400·	400·370	400·741	401·111	401·481	401·852	402·222	402·593	402·963	403·333	108
109	403·704	404·074	404·444	404·815	405·185	405·556	405·926	406·296	406·667	407·037	109
110	407·407	407·778	408·148	408·519	408·889	409·259	409·630	410·	410·370	410·741	110
111	411·111	411·481	411·852	412·222	412·593	412·963	413·333	413·704	414·074	414·444	111
112	414·815	415·185	415·556	415·926	416·296	416·667	417·037	417·407	417·778	418·148	112
113	418·519	418·889	419·259	419·630	420·	420·370	420·741	421·111	421·481	421·852	113
114	422·222	422·593	422·963	423·333	423·704	424·074	424·444	424·815	425·185	425·556	114
115	425·926	426·296	426·667	427·037	427·407	427·778	428·148	428·519	428·889	429·259	115
116	429·630	430·	430·370	430·741	431·111	431·481	431·852	432·222	432·593	432·963	116
117	433·333	433·704	434·074	434·444	434·815	435·185	435·556	435·926	436·296	436·667	117
118	437·037	437·407	437·778	438·148	438·519	438·889	439·259	439·630	440·	440·370	118
119	440·741	441·111	441·481	441·852	442·222	442·593	442·963	443·333	443·704	444·074	119
120	444·444	444·815	445·185	445·556	445·926	446·296	446·667	447·037	447·407	447·778	120
M.A.	·0	·1	·2	·3	·4	·5	6	·7	·8	·9	M.A.

CUBIC YARDS TO MEAN AREAS FOR 100 FEET IN LENGTH.

M.A.	·0	·1	·2	·3	·4	·5	·6	·7	·8	·9	M.A.
121	448·148	448·519	448·889	449·259	449·630	450·	450·370	450·741	451·111	451·481	121
122	451·852	452·222	452·593	452·963	453·333	453·704	454·074	454·444	454·815	455·185	122
123	455·556	455·926	456·296	456·667	457·037	457·407	457·778	458·148	458·519	458·889	123
124	459·259	459·630	460·	460·370	460·741	461·111	461·481	461·852	462·222	462·593	124
125	462·963	463·333	463·704	464·074	464·444	464·815	465·185	465·556	465·926	466·296	125
126	466·667	467·037	467·407	467·778	468·148	468·519	468·889	469·259	469·630	470·	126
127	470·370	470·741	471·111	471·481	471·852	472·222	472·593	472·963	473·333	473·704	127
128	474·074	474·444	474·815	475·185	475·556	475·926	476·296	476·667	477·037	477·407	128
129	477·778	478·148	478·519	478·889	479·259	479·630	480·	480·370	480·741	481·111	129
130	481·481	481·852	482·222	482·593	482·963	483·333	483·704	484·074	484·444	484·815	130
131	485·185	485·556	485·926	486·296	486·667	487·037	487·407	487·778	488·148	488·519	131
132	488·889	489·259	489·630	490·	490·370	490·741	491·111	491·481	491·852	492·222	132
133	492·593	492·963	493·333	493·704	494·074	494·444	494·815	495·185	495·556	495·926	133
134	496·296	496·667	497·037	497·407	497·778	498·148	498·519	498·889	499·259	499·630	134
135	500·	500·370	500·741	501·111	501·481	501·852	502·222	502·593	502·963	503·333	135
136	503·704	504·074	504·444	504·815	505·185	505·556	505·926	506·296	506·667	507·037	136
137	507·407	507·778	508·148	508·519	508·889	509·259	509·630	510·	510·370	510·741	137
138	511·111	511·481	511·852	512·222	512·593	512·963	513·333	513·704	514·074	514·444	138
139	514·815	515·185	515·556	515·926	516·296	516·667	517·037	517·407	517·778	518·148	139
140	518·519	518·889	519·259	519·630	520·	520·370	520·741	521·111	521·481	521·852	140
141	522·222	522·593	522·963	523·333	523·704	524·074	524·444	524·815	525·185	525·556	141
142	525·926	526·296	526·667	527·037	527·407	527·778	528·148	528·519	528·889	529·259	142
143	529·630	530·	530·370	530·741	531·111	531·481	531·852	532·222	532·593	532·963	143
144	533·333	533·704	534·074	534·444	534·815	535·185	535·556	535·926	536·296	536·667	144
145	537·037	537·407	537·778	538·148	538·519	538·889	539·259	539·630	540·	540·370	145
146	540·741	541·111	541·481	541·852	542·222	542·593	542·963	543·333	543·704	544·074	146
147	544·444	544·815	545·185	545·556	545·926	546·296	546·667	547·037	547·407	547·778	147
148	548·148	548·519	548·889	549·259	549·630	550·	550·370	550·741	551·111	551·481	148
149	551·852	552·222	552·593	552·963	553·333	553·704	554·074	554·444	554·815	555·185	149
150	555·556	555·926	556·296	556·667	557·037	557·407	557·778	558·148	558·519	558·889	150
151	559·259	559·630	560·	560·370	560·741	561·111	561·481	561·852	562·222	562·593	151
152	562·963	563·333	563·704	564·074	564·444	564·815	565·185	565·556	565·926	566·296	152
153	566·667	567·037	567·407	567·778	568·148	568·519	568·889	569·259	569·630	570·	153
154	570·370	570·741	571·111	571·481	571·852	572·222	572·593	572·963	573·333	573·704	154
155	574·074	574·444	574·815	575·185	575·556	575·926	576·296	576·667	577·037	577·407	155
156	577·778	578·148	578·519	578·889	579·259	579·630	580·	580·370	580·741	581·111	156
157	581·481	581·852	582·222	582·593	582·963	583·333	583·704	584·074	584·444	584·815	157
158	585·185	585·556	585·926	586·296	586·667	587·037	587·407	587·778	588·148	588·519	158
159	588·889	589·259	589·630	590·	590·370	590·741	591·111	591·481	591·852	592·222	159
160	592·593	592·963	593·333	593·704	594·074	594·444	594·815	595·185	595·556	595·926	160
161	596·296	596·667	597·037	597·407	597·778	598·148	598·519	598·889	599·259	599·630	161
162	600·	600·370	600·741	601·111	601·481	601·852	602·222	602·593	602·963	603·333	162
163	603·704	604·074	604·444	604·815	605·185	605·556	605·926	606·296	606·667	607·037	163
164	607·407	607·778	608·148	608·519	608·889	609·259	609·630	610·	610·370	610·741	164
165	611·111	611·481	611·852	612·222	612·593	612·963	613·333	613·704	614·074	614·444	165
166	614·815	615·185	615·556	615·926	616·296	616·667	617·037	617·407	617·778	618·148	166
167	618·519	618·889	619·259	619·630	620·	620·370	620·741	621·111	621·481	621·852	167
168	622·222	622·593	622·963	623·333	623·704	624·074	624·444	624·815	625·185	625·556	168
169	625·926	626·296	626·667	627·037	627·407	627·778	628·148	628·519	628·889	629·259	169
170	629·630	630·	630·370	630·741	631·111	631·481	631·852	632·222	632·593	632·963	170
171	633·333	633·704	634·074	634·444	634·815	635·185	635·556	635·926	636·296	636·667	171
172	637·037	637·407	637·778	638·148	638·519	638·889	639·259	639·630	640·	640·370	172
173	640·741	641·111	641·481	641·852	642·222	642·593	642·963	643·333	643·704	644·074	173
174	644·444	644·815	645·185	645·556	645·926	646·296	646·667	647·037	647·407	647·778	174
175	648·148	648·519	648·889	649·259	649·630	650·	650·370	650·741	651·111	651·481	175
176	651·852	652·222	652·593	652·963	653·333	653·704	654·074	654·444	654·815	655·185	176
177	655·556	655·926	656·296	656·667	657·037	657·407	657·778	658·148	658·519	658·889	177
178	659·259	659·630	660·	660·370	660·741	661·111	661·481	661·852	662·222	662·593	178
179	662·963	663·333	663·704	664·074	664·444	664·815	665·185	665·556	665·926	666·296	179
180	666·667	667·037	667·407	667·778	668·148	668·519	668·889	669·259	669·630	670·	180
M.A.	·0	·1	·2	·3	·4	·5	·6	·7	·8	·9	M.A.

MEAN AREAS 121 to 180.

CUBIC YARDS TO MEAN AREAS FOR 100 FEET IN LENGTH.

M.A.	·0	·1	·2	·3	·4	·5	·6	·7	·8	·9	M.A.
181	670·370	670·741	671·111	671·481	671·852	672·222	672·593	672·963	673·333	673·704	181
182	674·074	674·444	674·815	675·185	675·556	675·926	676·296	676·667	677·037	677·407	182
183	677·778	678·148	678·519	678·889	679·259	679·630	680·	680·370	680·741	681·111	183
184	681·481	681·852	682·222	682·593	682·963	6·3·333	683·704	684·074	684·444	684·815	184
185	685·185	685·556	685·926	686·296	686·667	687·037	687·407	687·778	688·148	688·519	185
186	688·889	689·259	689·630	690·	690·370	690·741	691·111	691·481	691·852	692·222	186
187	692·593	692·963	693·333	693·704	694·074	694·444	694·815	695·185	695·556	695·926	187
188	696·296	696·667	697·037	697·407	697·778	698·148	698·519	698·889	699·259	699·630	188
189	700·	700·370	700·741	701·111	701·481	701·852	702·222	702·593	702·963	703·333	189
190	703·704	704·074	704·444	704·815	705·185	705·556	705·926	706·296	706·667	707·037	190
191	707·407	707·778	708·148	708·519	708·889	709·259	709·630	710·	710·370	710·741	191
192	711·111	711·481	711·852	712·222	712·593	712·963	713·333	713·704	714·074	714·444	192
193	714·815	715·185	715·556	715·926	716·296	716·667	717·037	717·407	717·778	718·148	193
194	718·519	718·889	719·259	719·630	720·	720·370	720·741	721·111	721·481	721·852	194
195	722·222	722·593	722·963	723·333	723·704	724·074	724·444	724·815	725·185	725·556	195
196	725·926	726·296	726·667	727·037	727·407	727·778	728·148	728·519	728·889	729·259	196
197	729·630	730·	730·370	730·741	731·111	731·481	731·852	732·222	732·593	732·963	197
198	733·333	733·704	734·074	734·444	734·815	735·185	735·556	735·926	736·296	736·667	198
199	737·037	737·407	737·778	738·148	738·519	738·889	739·259	739·630	740·	740·370	199
200	740·741	741·111	741·481	741·852	742·222	742·593	742·963	743·333	743·704	744·074	200
201	744·444	744·815	745·185	745·556	745·926	746·296	746·667	747·037	747·407	747·778	201
202	748·148	748·519	748·889	749·259	749·630	750·	750·370	750·741	751·111	751·481	202
203	751·852	752·222	752·593	752·963	753·333	753·704	754·074	754·444	754·815	755·185	203
204	755·556	755·926	756·296	756·667	757·037	757·407	757·778	758·148	758·519	758·889	204
205	759·259	759·630	760·	760·370	760·741	761·111	761·481	761·852	762·222	762·593	205
206	762·963	763·333	763·704	764·074	764·444	764·815	765·185	765·556	765·926	766·296	206
207	766·667	767·037	767·407	767·778	768·148	768·519	768·889	769·259	769·630	770·	207
208	770·370	770·741	771·111	771·481	771·852	772·222	772·593	772·963	773·333	773·704	208
209	774·074	774·444	774·815	775·185	775·556	775·926	776·296	776·667	777·037	777·407	209
210	777·778	778·148	778·519	778·889	779·259	779·630	780·	780·370	780·741	781·111	210
211	781·481	781·852	782·222	782·593	782·963	783·333	783·704	784·074	784·444	784·815	211
212	785·185	785·556	785·926	786·296	786·667	787·037	787·407	787·778	788·148	788·519	212
213	788·889	789·259	789·630	790·	790·370	790·741	791·111	791·481	791·852	792·222	213
214	792·593	792·963	793·333	793·704	794·074	794·444	794·815	795·185	795·556	795·926	214
215	796·296	796·667	797·037	797·407	797·778	798·148	798·519	798·889	799·259	799·630	215
216	800·	800·370	800·741	801·111	801·481	801·852	802·222	802·593	802·963	803·333	216
217	803·704	804·074	804·444	804·815	805·185	805·556	805·926	806·296	806·667	807·037	217
218	807·407	807·778	808·148	808·519	808·889	809·259	809·630	810·	810·370	810·741	218
219	811·111	811·481	811·852	812·222	812·593	812·963	813·333	813·704	814·074	814·444	219
220	814·815	815·185	815·556	815·926	816·296	816·667	817·037	817·407	817·778	818·148	220
221	818·519	818·889	819·259	819·630	820·	820·370	820·741	821·111	821·481	821·852	221
222	822·222	822·593	822·963	823·333	823·704	824·074	824·444	824·815	825·185	825·556	222
223	825·926	826·296	826·667	827·037	827·407	827·778	828·148	828·519	828·889	829·259	223
224	829·630	830·	830·370	830·741	831·111	831·481	831·852	832·222	832·593	832·963	224
225	833·333	833·704	834·074	834·444	834·815	835·185	835·556	835·926	836·296	836·667	225
226	837·037	837·407	837·778	838·148	838·519	838·889	839·259	839·630	840·	840·370	226
227	840·741	841·111	841·481	841·852	842·222	842·593	842·963	843·333	843·704	844·074	227
228	844·444	844·815	845·185	845·556	845·926	846·296	846·667	847·037	847·407	847·778	228
229	848·148	848·519	848·889	849·259	849·630	850·	850·370	850·741	851·111	851·481	229
230	851·852	852·222	852·593	852·963	853·333	853·704	854·074	854·444	854·815	855·185	230
231	855·556	855·926	856·296	856·667	857·037	857·407	857·778	858·148	858·519	858·889	231
232	859·259	859·630	860·	860·370	860·741	861·111	861·481	861·852	862·222	862·593	232
233	862·963	863·333	863·704	864·074	864·444	864·815	865·185	865·556	865·926	866·296	233
234	866·667	867·037	867·407	867·778	868·148	868·519	868·889	869·259	869·630	870·	234
235	870·370	870·741	871·111	871·481	871·852	872·222	872·593	872·963	873·333	873·704	235
236	874·074	874·444	874·815	875·185	875·556	875·926	876·296	876·667	877·037	877·407	236
237	877·778	878·148	878·519	878·889	879·259	879·630	880·	880·370	880·741	881·111	237
238	881·481	881·852	882·222	882·593	882·963	883·333	883·704	884·074	884·444	884·815	238
239	885·185	885·556	885·926	886·296	886·667	887·037	887·407	887·778	888·148	888·519	239
240	888·889	889·259	889·630	890·	890·370	890·741	891·111	891·481	891·852	892·222	240
M.A.	·0	·1	·2	·3	·4	·5	·6	·7	·8	·9	M.A.

MEAN AREAS 181 to 240.

CUBIC YARDS TO MEAN AREAS FOR 100 FEET IN LENGTH.

M.A.	·0	·1	·2	·3	·4	·5	·6	·7	·8	·9	M.A.
241	892·593	892·963	893·333	893·704	894·074	894·444	894·815	895·185	895·556	895·926	241
242	896·296	896·667	897·037	897·407	897·778	898·148	898·519	898·889	899·259	899·630	242
243	900·	900·370	900·741	901·111	901·481	901·852	902·222	902·593	902·963	903·333	243
244	903·704	904·074	904·444	904·815	905·185	905·556	905·926	906·296	906·667	907·037	244
245	907·407	907·778	908·148	908·519	908·889	909·259	909·630	910·	910·370	910·741	245
246	911·111	911·481	911·852	912·222	912·593	912·963	913·333	913·704	914·074	914·444	246
247	914·815	915·185	915·556	915·926	916·296	916·667	917·037	917·407	917·778	918·148	247
248	918·519	918·889	919·259	919·630	920·	920·370	920·741	921·111	921·481	921·852	248
249	922·222	922·593	922·963	923·333	923·704	924·074	924·444	924·815	925·185	925·556	249
250	925·926	926·296	926·667	927·037	927·407	927·778	928·148	928·519	928·889	929·259	250
251	929·630	930·	930·370	930·741	931·111	931·481	931·852	932·222	932·593	932·963	251
252	933·333	933·704	934·074	934·444	934·815	935·185	935·556	935·926	936·296	936·667	252
253	937·037	937·407	937·778	938·148	938·519	938·889	939·259	939·630	940·	940·370	253
254	940·741	941·111	941·481	941·852	942·222	942·593	942·963	943·333	943·704	944·074	254
255	944·444	944·815	945·185	945·556	945·926	946·296	946·667	947·037	947·407	947·778	255
256	948·148	948·519	948·889	949·259	949·630	950·	950·370	950·741	951·111	951·481	256
257	951·852	952·222	952·593	952·963	953·333	953·704	954·074	954·444	954·815	955·185	257
258	955·556	955·926	956·296	956·667	957·037	957·407	957·778	958·148	958·519	958·889	258
259	959·259	959·630	960·	960·370	960·741	961·111	961·481	961·852	962·222	962·593	259
260	962·963	963·333	963·704	964·074	964·444	964·815	965·185	965·556	965·926	966·296	260
261	966·667	967·037	967·407	967·778	968·148	968·519	968·889	969·259	969·630	970·	261
262	970·370	970·741	971·111	971·481	971·852	972·222	972·593	972·963	973·333	973·704	262
263	974·074	974·444	974·815	975·185	975·556	975·926	976·296	976·667	977·037	977·407	263
264	977·778	978·148	978·519	978·889	979·259	979·630	980·	980·370	980·741	981·111	264
265	981·481	981·852	982·222	982·593	982·963	983·333	983·704	984·074	984·444	984·815	265
266	985·185	985·556	985·926	986·296	986·667	987·037	987·407	987·778	988·148	988·519	266
267	988·889	989·259	989·630	990·	990·370	990·741	991·111	991·481	991·852	992·222	267
268	992·593	992·963	993·333	993·704	994·074	994·444	994·815	995·185	995·556	995·926	268
269	996·296	996·667	997·037	997·407	997·778	998·148	998·519	998·889	999·259	999·630	269
270	1000·	1000·370	1000·741	1001·111	1001·481	1001·852	1002·222	1002·593	1002·963	1003·333	270
271	1003·704	1004·074	1004·444	1004·815	1005·185	1005·556	1005·926	1006·296	1006·667	1007·037	271
272	1007·407	1007·778	1008·148	1008·519	1008·889	1009·259	1009·630	1010·	1010·370	1010·741	272
273	1011·111	1011·481	1011·852	1012·222	1012·593	1012·963	1013·333	1013·704	1014·074	1014·444	273
274	1014·815	1015·185	1015·556	1015·926	1016·296	1016·667	1017·037	1017·407	1017·778	1018·148	274
275	1018·519	1018·889	1019·259	1019·630	1020·	1020·370	1020·741	1021·111	1021·481	1021·852	275
276	1022·222	1022·593	1022·963	1023·333	1023·704	1024·074	1024·444	1024·815	1025·185	1025·556	276
277	1025·926	1026·296	1026·667	1027·037	1027·407	1027·778	1028·148	1028·519	1028·889	1029·259	277
278	1029·630	1030·	1030·370	1030·741	1031·111	1031·481	1031·852	1032·222	1032·593	1032·963	278
279	1033·333	1033·704	1034·074	1034·444	1034·815	1035·185	1035·556	1035·926	1036·296	1036·667	279
280	1037·037	1037·407	1037·778	1038·148	1038·519	1038·889	1039·259	1039·630	1040·	1040·370	280
281	1040·741	1041·111	1041·481	1041·852	1042·222	1042·593	1042·963	1043·333	1043·704	1044·074	281
282	1044·444	1044·815	1045·185	1045·556	1045·926	1046·296	1046·667	1047·037	1047·407	1047·778	282
283	1048·148	1048·519	1048·889	1049·259	1049·630	1050·	1050·370	1050·741	1051·111	1051·481	283
284	1051·852	1052·222	1052·593	1052·963	1053·333	1053·704	1054·074	1054·444	1054·815	1055·185	284
285	1055·556	1055·926	1056·296	1056·667	1057·037	1057·407	1057·778	1058·148	1058·519	1058·889	285
286	1059·259	1059·630	1060·	1060·370	1060·741	1061·111	1061·481	1061·852	1062·222	1062·593	286
287	1062·963	1063·333	1063·704	1064·074	1064·444	1064·815	1065·185	1065·556	1065·926	1066·296	287
288	1066·667	1067·037	1067·407	1067·778	1068·148	1068·519	1068·889	1069·259	1069·630	1070·	288
289	1070·370	1070·741	1071·111	1071·481	1071·852	1072·222	1072·593	1072·963	1073·333	1073·704	289
290	1074·074	1074·444	1074·815	1075·185	1075·556	1075·926	1076·296	1076·667	1077·037	1077·407	290
291	1077·778	1078·148	1078·519	1078·889	1079·259	1079·630	1080·	1080·370	1080·741	1081·111	291
292	1081·481	1081·852	1082·222	1082·593	1082·963	1083·333	1083·704	1084·074	1084·444	1084·815	292
293	1085·185	1085·556	1085·926	1086·296	1086·667	1087·037	1087·407	1087·778	1088·148	1088·519	293
294	1088·889	1089·259	1089·630	1090·	1090·370	1090·741	1091·111	1091·481	1091·852	1092·222	294
295	1092·593	1092·963	1093·333	1093·704	1094·074	1094·444	1094·815	1095·185	1095·556	1095·926	295
296	1096·296	1096·667	1097·037	1097·407	1097·778	1098·148	1098·519	1098·889	1099·259	1099·630	296
297	1100·	1100·370	1100·741	1101·111	1101·481	1101·852	1102·222	1102·593	1102·963	1103·333	297
298	1103·704	1104·074	1104·444	1104·815	1105·185	1105·556	1105·926	1106·296	1106·667	1107·037	298
299	1107·407	1107·778	1108·148	1108·519	1108·889	1109·259	1109·630	1110·	1110·370	1110·741	299
300	1111·111	1111·481	1111·852	1112·222	1112·593	1112·963	1113·333	1113·704	1114·074	1114·444	300
M.A.	·0	·1	·2	·3	·4	·5	·6	·7	·8	·9	M.A.

MEAN AREAS 241 to 300.

CUBIC YARDS TO MEAN AREAS FOR 100 FEET IN LENGTH.

M.A.	·0	·1	·2	·3	·4	·5	·6	·7	·8	·9	M.A.
301	1114·815	1115·185	1115·556	1115·926	1116·296	1116·667	1117·037	1117·407	1117·778	1118·148	301
302	1118·519	1118·889	1119·259	1119·630	1120·	1120·370	1120·741	1121·111	1121·481	1121·852	302
303	1122·222	1122·593	1122·963	1123·333	1123·704	1124·074	1124·444	1124·815	1125·185	1125·556	303
304	1125·926	1126·296	1126·667	1127·037	1127·407	1127·778	1128·148	1128·519	1128·889	1129·259	304
305	1129·630	1130·	1130·370	1130·741	1131·111	1131·481	1131·852	1132·222	1132·593	1132·963	305
306	1133·333	1133·704	1134·074	1134·444	1134·815	1135·185	1135·556	1135·926	1136·296	1136·667	306
307	1137·037	1137·407	1137·778	1138·148	1138·519	1138·889	1139·259	1139·630	1140·	1140·370	307
308	1140·741	1141·111	1141·481	1141·852	1142·222	1142·593	1142·963	1143·333	1143·704	1144·074	308
309	1144·444	1144·815	1145·185	1145·556	1145·926	1146·296	1146·667	1147·037	1147·407	1147·778	309
310	1148·148	1148·519	1148·889	1149·259	1149·630	1150·	1150·370	1150·741	1151·111	1151·481	310
311	1151·852	1152·222	1152·593	1152·963	1153·333	1153·704	1154·074	1154·444	1154·815	1155·185	311
312	1155·556	1155·926	1156·296	1156·667	1157·037	1157·407	1157·778	1158·148	1158·519	1158·889	312
313	1159·259	1159·630	1160·	1160·370	1160·741	1161·111	1161·481	1161·852	1162·222	1162·593	313
314	1162·963	1163·333	1163·704	1164·074	1164·444	1164·815	1165·185	1165·556	1165·926	1166·296	314
315	1166·667	1167·037	1167·407	1167·778	1168·148	1168·519	1168·889	1169·259	1169·630	1170·	315
316	1170·370	1170·741	1171·111	1171·481	1171·852	1172·222	1172·593	1172·963	1173·333	1173·704	316
317	1174·074	1174·444	1174·815	1175·185	1175·556	1175·926	1176·296	1176·667	1177·037	1177·407	317
318	1177·778	1178·148	1178·519	1178·889	1179·259	1179·630	1180·	1180·370	1180·741	1181·111	318
319	1181·481	1181·852	1182·222	1182·593	1182·963	1183·333	1183·704	1184·074	1184·444	1184·815	319
320	1185·185	1185·556	1185·926	1186·296	1186·667	1187·037	1187·407	1187·778	1188·148	1188·519	320
321	1188·889	1189·259	1189·630	1190·	1190·370	1190·741	1191·111	1191·481	1191·852	1192·222	321
322	1192·593	1192·963	1193·333	1193·704	1194·074	1194·444	1194·815	1195·185	1195·556	1195·926	322
323	1196·296	1196·667	1197·037	1197·407	1197·778	1198·148	1198·519	1198·889	1199·259	1199·630	323
324	1200·	1200·370	1200·741	1201·111	1201·481	1201·852	1202·222	1202·593	1202·963	1203·333	324
325	1203·704	1204·074	1204·444	1204·815	1205·185	1205·556	1205·926	1206·296	1206·667	1207·037	325
326	1207·407	1207·778	1208·148	1208·519	1208·889	1209·259	1209·630	1210·	1210·370	1210·741	326
327	1211·111	1211·481	1211·852	1212·222	1212·593	1212·963	1213·333	1213·704	1214·074	1214·444	327
328	1214·815	1215·185	1215·556	1215·926	1216·296	1216·667	1217·037	1217·407	1217·778	1218·148	328
329	1218·519	1218·889	1219·259	1219·630	1220·	1220·370	1220·741	1221·111	1221·481	1221·852	329
330	1222·222	1222·593	1222·963	1223·333	1223·704	1224·074	1224·444	1224·815	1225·185	1225·556	330
331	1225·926	1226·296	1226·667	1227·037	1227·407	1227·778	1228·148	1228·519	1228·889	1229·259	331
332	1229·630	1230·	1230·370	1230·741	1231·111	1231·481	1231·852	1232·222	1232·593	1232·963	332
333	1233·333	1233·704	1234·074	1234·444	1234·815	1235·185	1235·556	1235·926	1236·296	1236·667	333
334	1237·037	1237·407	1237·778	1238·148	1238·519	1238·889	1239·259	1239·630	1240·	1240·370	334
335	1240·741	1241·111	1241·481	1241·852	1242·222	1242·593	1242·963	1243·333	1243·704	1244·074	335
336	1244·444	1244·815	1245·185	1245·556	1245·926	1246·296	1246·667	1247·037	1247·407	1247·778	336
337	1248·148	1248·519	1248·889	1249·259	1249·630	1250·	1250·370	1250·741	1251·111	1251·481	337
338	1251·852	1252·222	1252·593	1252·963	1253·333	1253·704	1254·074	1254·444	1254·815	1255·185	338
339	1255·556	1255·926	1256·296	1256·667	1257·037	1257·407	1257·778	1258·148	1258·519	1258·889	339
340	1259·259	1259·630	1260·	1260·370	1260·741	1261·111	1261·481	1261·852	1262·222	1262·593	340
341	1262·963	1263·333	1263·704	1264·074	1264·444	1264·815	1265·185	1265·556	1265·926	1266·296	341
342	1266·667	1267·037	1267·407	1267·778	1268·148	1268·519	1268·889	1269·259	1269·630	1270·	342
343	1270·370	1270·741	1271·111	1271·481	1271·852	1272·222	1272·593	1272·963	1273·333	1273·704	343
344	1274·074	1274·444	1274·815	1275·185	1275·556	1275·926	1276·296	1276·667	1277·037	1277·407	344
345	1277·778	1278·148	1278·519	1278·889	1279·259	1279·630	1280·	1280·370	1280·741	1281·111	345
346	1281·481	1281·852	1282·222	1282·593	1282·963	1283·333	1283·704	1284·074	1284·444	1284·815	346
347	1285·185	1285·556	1285·926	1286·296	1286·667	1287·037	1287·407	1287·778	1288·148	1288·519	347
348	1288·889	1289·259	1289·630	1290·	1290·370	1290·741	1291·111	1291·481	1291·852	1292·222	348
349	1292·593	1292·963	1293·333	1293·704	1294·074	1294·444	1294·815	1295·185	1295·556	1295·926	349
350	1296·296	1296·667	1297·037	1297·407	1297·778	1298·148	1298·519	1298·889	1299·259	1299·630	350
351	1300·	1300·370	1300·741	1301·111	1301·481	1301·852	1302·222	1302·593	1302·963	1303·333	351
352	1303·704	1304·074	1304·444	1304·815	1305·185	1305·556	1305·926	1306·296	1306·667	1307·037	352
353	1307·407	1307·778	1308·148	1308·519	1308·889	1309·259	1309·630	1310·	1310·370	1310·741	353
354	1311·111	1311·481	1311·852	1312·222	1312·593	1312·963	1313·333	1313·704	1314·074	1314·444	354
355	1314·815	1315·185	1315·556	1315·926	1316·296	1316·667	1317·037	1317·407	1317·778	1318·148	355
356	1318·519	1318·889	1319·259	1319·630	1320·	1320·370	1320·741	1321·111	1321·481	1321·852	356
357	1322·222	1322·593	1322·963	1323·333	1323·704	1324·074	1324·444	1324·815	1325·185	1325·556	357
358	1325·926	1326·296	1326·667	1327·037	1327·407	1327·778	1328·148	1328·519	1328·889	1329·259	358
359	1329·630	1330·	1330·370	1330·741	1331·111	1331·481	1331·852	1332·222	1332·593	1332·963	359
360	1333·333	1333·704	1334·074	1334·444	1334·815	1335·185	1335·556	1335·926	1336·296	1336·667	360
M.A.	·0	·1	·2	·3	·4	·5	·6	·7	·8	·9	M.A.

MEAN AREAS 301 to 360.

CUBIC YARDS TO MEAN AREAS FOR 100 FEET IN LENGTH.

M.A.	·0	·1	·2	·3	·4	·5	·6	·7	·8	·9	M.A.
361	1337·037	1337·407	1337·778	1338·148	1338·519	1338·889	1339·259	1339·630	1340·	1340·370	361
362	1340·741	1341·111	1341·481	1341·852	1342·222	1342·593	1342·963	1343·333	1343·704	1344·074	362
363	1344·444	1344·815	1345·185	1345·556	1345·926	1346·296	1346·667	1347·037	1347·407	1347·778	363
364	1348·148	1348·519	1348·889	1349·259	1349·630	1350·	1350·370	1350·741	1351·111	1351·481	364
365	1351·852	1352·222	1352·593	1352·963	1353·333	1353·704	1354·074	1354·444	1354·815	1355·185	365
366	1355·556	1355·926	1356·296	1356·667	1357·037	1357·407	1357·778	1358·148	1358·519	1358·889	366
367	1359·259	1359·630	1360·	1360·370	1360·741	1361·111	1361·481	1361·852	1362·222	1362·593	367
368	1362·963	1363·333	1363·704	1364·074	1364·444	1364·815	1365·185	1365·556	1365·926	1366·296	368
369	1366·667	1367·037	1367·407	1367·778	1368·148	1368·519	1368·889	1369·259	1369·630	1370·	369
370	1370·370	1370·741	1371·111	1371·481	1371·852	1372·222	1372·593	1372·963	1373·333	1373·704	370
371	1374·074	1374·444	1374·815	1375·185	1375·556	1375·926	1376·296	1376·667	1377·037	1377·407	371
372	1377·778	1378·148	1378·519	1378·889	1379·259	1379·630	1380·	1380·370	1380·741	1381·111	372
373	1381·481	1381·852	1382·222	1382·593	1382·963	1383·333	1383·704	1384·074	1384·444	1384·815	373
374	1385·185	1385·556	1385·926	1386·296	1386·667	1387·037	1387·407	1387·778	1388·148	1388·519	374
375	1388·889	1389·259	1389·630	1390·	1390·370	1390·741	1391·111	1391·481	1391·852	1392·222	375
376	1392·593	1392·963	1393·333	1393·704	1394·074	1394·444	1394·815	1395·185	1395·556	1395·926	376
377	1396·296	1396·667	1397·037	1397·407	1397·778	1398·148	1398·519	1398·889	1399·259	1399·630	377
378	1400·	1400·370	1400·741	1401·111	1401·481	1401·852	1402·222	1402·593	1402·963	1403·333	378
379	1403·704	1404·074	1404·444	1404·815	1405·185	1405·556	1405·926	1406·296	1406·667	1407·037	379
380	1407·407	1407·778	1408·148	1408·519	1408·889	1409·259	1409·630	1410·	1410·370	1410·741	380
381	1411·111	1411·481	1411·852	1412·222	1412·593	1412·963	1413·333	1413·704	1414·074	1414·444	381
382	1414·815	1415·185	1415·556	1415·926	1416·296	1416·667	1417·037	1417·407	1417·778	1418·148	382
383	1418·519	1418·889	1419·259	1419·630	1420·	1420·370	1420·741	1421·111	1421·481	1421·852	383
384	1422·222	1422·593	1422·963	1423·333	1423·704	1424·074	1424·444	1424·815	1425·185	1425·556	384
385	1425·926	1426·296	1426·667	1427·037	1427·407	1427·778	1428·148	1428·519	1428·889	1429·259	385
386	1429·630	1430·	1430·370	1430·741	1431·111	1431·481	1431·852	1432·222	1432·593	1432·963	386
387	1433·333	1433·704	1434·074	1434·444	1434·815	1435·185	1435·556	1435·926	1436·296	1436·667	387
388	1437·037	1437·407	1437·778	1438·148	1438·519	1438·889	1439·259	1439·630	1440·	1440·370	388
389	1440·741	1441·111	1441·481	1441·852	1442·222	1442·593	1442·963	1443·333	1443·704	1444·074	389
390	1444·444	1444·815	1445·185	1445·556	1445·926	1446·296	1446·667	1447·037	1447·407	1447·778	390
391	1448·148	1448·519	1448·889	1449·259	1449·630	1450·	1450·370	1450·741	1451·111	1451·481	391
392	1451·852	1452·222	1452·593	1452·963	1453·333	1453·704	1454·074	1454·444	1454·815	1455·185	392
393	1455·556	1455·926	1456·296	1456·667	1457·037	1457·407	1457·778	1458·148	1458·519	1458·889	393
394	1459·259	1459·630	1460·	1460·370	1460·741	1461·111	1461·481	1461·852	1462·222	1462·593	394
395	1462·963	1463·333	1463·704	1464·074	1464·444	1464·815	1465·185	1465·556	1465·926	1466·296	395
396	1466·667	1467·037	1467·407	1467·778	1468·148	1468·519	1468·889	1469·259	1469·630	1470·	396
397	1470·370	1470·741	1471·111	1471·481	1471·852	1472·222	1472·593	1472·963	1473·333	1473·704	397
398	1474·074	1474·444	1474·815	1475·185	1475·556	1475·926	1476·296	1476·667	1477·037	1477·407	398
399	1477·778	1478·148	1478·519	1478·889	1479·259	1479·630	1480·	1480·370	1480·741	1481·111	399
400	1481·481	1481·852	1482·222	1482·593	1482·963	1483·333	1483·704	1484·074	1484·444	1484·815	400
401	1485·185	1485·556	1485·926	1486·296	1486·667	1487·037	1487·407	1487·778	1488·148	1488·519	401
402	1488·889	1489·259	1489·630	1490·	1490·370	1490·741	1491·111	1491·481	1491·852	1492·222	402
403	1492·593	1492·963	1493·333	1493·704	1494·074	1494·444	1494·815	1495·185	1495·556	1495·926	403
404	1496·296	1496·667	1497·037	1497·407	1497·778	1498·148	1498·519	1498·889	1499·259	1499·630	404
405	1500·	1500·370	1500·741	1501·111	1501·481	1501·852	1502·222	1502·593	1502·963	1503·333	405
406	1503·704	1504·074	1504·444	1504·815	1505·185	1505·556	1505·926	1506·296	1506·667	1507·037	406
407	1507·407	1507·778	1508·148	1508·519	1508·889	1509·259	1509·630	1510·	1510·370	1510·741	407
408	1511·111	1511·481	1511·852	1512·222	1512·593	1512·963	1513·333	1513·704	1514·074	1514·444	408
409	1514·815	1515·185	1515·556	1515·926	1516·296	1516·667	1517·037	1517·407	1517·778	1518·148	409
410	1518·519	1518·889	1519·259	1519·630	1520·	1520·370	1520·741	1521·111	1521·481	1521·852	410
411	1522·222	1522·593	1522·963	1523·333	1523·704	1524·074	1524·444	1524·815	1525·185	1525·556	411
412	1525·926	1526·296	1526·667	1527·037	1527·407	1527·778	1528·148	1528·519	1528·889	1529·259	412
413	1529·630	1530·	1530·370	1530·741	1531·111	1531·481	1531·852	1532·222	1532·593	1532·963	413
414	1533·333	1533·704	1534·074	1534·444	1534·815	1535·185	1535·556	1535·926	1536·296	1536·667	414
415	1537·037	1537·407	1537·778	1538·148	1538·519	1538·889	1539·259	1539·630	1540·	1540·370	415
416	1540·741	1541·111	1541·481	1541·852	1542·222	1542·593	1542·963	1543·333	1543·704	1544·074	416
417	1544·444	1544·815	1545·185	1545·556	1545·926	1546·296	1546·667	1547·037	1547·407	1547·778	417
418	1548·148	1548·519	1548·889	1549·259	1549·630	1550·	1550·370	1550·741	1551·111	1551·481	418
419	1551·852	1552·222	1552·593	1552·963	1553·333	1553·704	1554·074	1554·444	1554·815	1555·185	419
420	1555·556	1555·926	1556·296	1556·667	1557·037	1557·407	1557·778	1558·148	1558·519	1558·889	420
M.A.	·0	·1	·2	·3	·4	·5	·6	·7	·8	·9	M.A.

MEAN AREAS 361 to 420.

CUBIC YARDS TO MEAN AREAS FOR 100 FEET IN LENGTH.

M.A.	·0	·1	·2	·3	·4	·5	·6	·7	·8	·9	M.A.
421	1559·250	1559·630	1560·	1560·370	1560·741	1561·111	1561·481	1561·852	1562·222	1562·593	421
422	1562·963	1563·333	1563·704	1564·074	1564·444	1564·815	1565·185	1565·556	1565·926	1566·296	422
423	1566·667	1567·037	1567·407	1567·778	1568·148	1568·519	1568·889	1569·259	1569·630	1570·	423
424	1570·370	1570·741	1571·111	1571·481	1571·852	1572·222	1572·593	1572·963	1573·333	1573·704	424
425	1574·074	1574·444	1574·815	1575·185	1575·556	1575·926	1576·296	1576·667	1577·037	1577·407	425
426	1577·778	1578·148	1578·519	1578·889	1579·259	1579·630	1580·	1580·370	1580·741	1581·111	426
427	1581·481	1581·852	1582·222	1582·593	1582·963	1583·333	1583·704	1584·074	1584·444	1584·815	427
428	1585·185	1585·556	1585·926	1586·296	1586·667	1587·037	1587·407	1587·778	1588·148	1588·519	428
429	1588·889	1589·259	1589·630	1590·	1590·370	1590·741	1591·111	1591·481	1591·852	1592·222	429
430	1592·593	1592·963	1593·333	1593·704	1594·074	1594·444	1594·815	1595·185	1595·556	1595·926	430
431	1596·296	1596·667	1597·037	1597·407	1597·778	1598·148	1598·519	1598·889	1599·259	1599·630	431
432	1600·	1600·370	1600·741	1601·111	1601·481	1601·852	1602·222	1602·593	1602·963	1603·333	432
433	1603·704	1604·074	1604·444	1604·815	1605·185	1605·556	1605·926	1606·296	1606·667	1607·037	433
434	1607·407	1607·778	1608·148	1608·519	1608·889	1609·259	1609·630	1610·	1610·370	1610·741	434
435	1611·111	1611·481	1611·852	1612·222	1612·593	1612·963	1613·333	1613·704	1614·074	1614·444	435
436	1614·815	1615·185	1615·556	1615·926	1616·296	1616·667	1617·037	1617·407	1617·778	1618·148	436
437	1618·519	1618·889	1619·259	1619·630	1620·	1620·370	1620·741	1621·111	1621·481	1621·852	437
438	1622·222	1622·593	1622·963	1623·333	1623·704	1624·074	1624·444	1624·815	1625·185	1625·556	438
439	1625·926	1626·296	1626·667	1627·037	1627·407	1627·778	1628·148	1628·519	1628·889	1629·259	439
440	1629·630	1630·	1630·370	1630·741	1631·111	1631·481	1631·852	1632·222	1632·593	1632·963	440
441	1633·333	1633·704	1634·074	1634·444	1634·815	1635·185	1635·556	1635·926	1636·296	1636·667	441
442	1637·037	1637·407	1637·778	1638·148	1638·519	1638·889	1639·259	1639·630	1640·	1640·370	442
443	1640·741	1641·111	1641·481	1641·852	1642·222	1642·593	1642·963	1643·333	1643·704	1644·074	443
444	1644·444	1644·815	1645·185	1645·556	1645·926	1646·296	1646·667	1647·037	1647·407	1647·778	444
445	1648·148	1648·519	1648·889	1649·259	1649·630	1650·	1650·370	1650·741	1651·111	1651·481	445
446	1651·852	1652·222	1652·593	1652·963	1653·333	1653·704	1654·074	1654·444	1654·815	1655·185	446
447	1655·556	1655·926	1656·296	1656·667	1657·037	1657·407	1657·778	1658·148	1658·519	1658·889	447
448	1659·259	1659·630	1660·	1660·370	1660·741	1661·111	1661·481	1661·852	1662·222	1662·593	448
449	1662·963	1663·333	1663·704	1664·074	1664·444	1664·815	1665·185	1665·556	1665·926	1666·296	449
450	1666·667	1667·037	1667·407	1667·778	1668·148	1668·519	1668·889	1669·259	1669·630	1670·	450
451	1670·370	1670·741	1671·111	1671·481	1671·852	1672·222	1672·593	1672·963	1673·333	1673·704	451
452	1674·074	1674·444	1674·815	1675·185	1675·556	1675·926	1676·296	1676·667	1677·037	1677·407	452
453	1677·778	1678·148	1678·519	1678·889	1679·259	1679·630	1680·	1680·370	1680·741	1681·111	453
454	1681·481	1681·852	1682·222	1682·593	1682·963	1683·333	1683·704	1684·074	1684·444	1684·815	454
455	1685·185	1685·556	1685·926	1686·296	1686·667	1687·037	1687·407	1687·778	1688·148	1688·519	455
456	1688·889	1689·259	1689·630	1690·	1690·370	1690·741	1691·111	1691·481	1691·852	1692·222	456
457	1692·593	1692·963	1693·333	1693·704	1694·074	1694·444	1694·815	1695·185	1695·556	1695·926	457
458	1696·296	1696·667	1697·037	1697·407	1697·778	1698·148	1698·519	1698·889	1699·259	1699·630	458
459	1700·	1700·370	1700·741	1701·111	1701·481	1701·852	1702·222	1702·593	1702·963	1703·333	459
460	1703·704	1704·074	1704·444	1704·815	1705·185	1705·556	1705·926	1706·296	1706·667	1707·037	460
461	1707·407	1707·778	1708·148	1708·519	1708·889	1709·259	1709·630	1710·	1710·370	1710·741	461
462	1711·111	1711·481	1711·852	1712·222	1712·593	1712·963	1713·333	1713·704	1714·074	1714·444	462
463	1714·815	1715·185	1715·556	1715·926	1716·296	1716·667	1717·037	1717·407	1717·778	1718·148	463
464	1718·519	1718·889	1719·259	1719·630	1720·	1720·370	1720·741	1721·111	1721·481	1721·852	464
465	1722·222	1722·593	1722·963	1723·333	1723·704	1724·074	1724·444	1724·815	1725·185	1725·556	465
466	1725·926	1726·296	1726·667	1727·037	1727·407	1727·778	1728·148	1728·519	1728·889	1729·259	466
467	1729·630	1730·	1730·370	1730·741	1731·111	1731·481	1731·852	1732·222	1732·593	1732·963	467
468	1733·333	1733·704	1734·074	1734·444	1734·815	1735·185	1735·556	1735·926	1736·296	1736·667	468
469	1737·037	1737·407	1737·778	1738·148	1738·519	1738·889	1739·259	1739·630	1740·	1740·370	469
470	1740·741	1741·111	1741·481	1741·852	1742·222	1742·593	1742·963	1743·333	1743·704	1744·074	470
471	1744·444	1744·815	1745·185	1745·556	1745·926	1746·296	1746·667	1747·037	1747·407	1747·778	471
472	1748·148	1748·519	1748·889	1749·259	1749·630	1750·	1750·370	1750·741	1751·111	1751·481	472
473	1751·852	1752·222	1752·593	1752·963	1753·333	1753·704	1754·074	1754·444	1754·815	1755·185	473
474	1755·556	1755·926	1756·296	1756·667	1757·037	1757·407	1757·778	1758·148	1758·519	1758·889	474
475	1759·259	1759·630	1760·	1760·370	1760·741	1761·111	1761·481	1761·852	1762·222	1762·593	475
476	1762·963	1763·333	1763·704	1764·074	1764·444	1764·815	1765·185	1765·556	1765·926	1766·296	476
477	1766·667	1767·037	1767·407	1767·778	1768·148	1768·519	1768·889	1769·259	1769·630	1770·	477
478	1770·370	1770·741	1771·111	1771·481	1771·852	1772·222	1772·593	1772·963	1773·333	1773·704	478
479	1774·074	1774·444	1774·815	1775·185	1775·556	1775·926	1776·296	1776·667	1777·037	1777·407	479
480	1777·778	1778·148	1778·519	1778·889	1779·259	1779·630	1780·	1780·370	1780·741	1781·111	480
M.A.	·0	·1	·2	·3	·4	·5	·6	·7	·8	·9	M.A.

MEAN AREAS 421 to 480.

CUBIC YARDS TO MEAN AREAS FOR 100 FEET IN LENGTH.

M.A.	·0	·1	·2	·3	·4	·5	·6	·7	·8	·9	M.A.
481	1781·481	1781·852	1782·222	1782·593	1782·963	1783·333	1783·704	1784·074	1784·444	1784·815	481
482	1785·185	1785·556	1785·926	1786·296	1786·667	1787·037	1787·407	1787·778	1788·148	1788·519	482
483	1788·889	1789·259	1789·630	1790·	1790·370	1790·741	1791·111	1791·481	1791·852	1792·222	483
484	1792·593	1792·963	1793·333	1793·704	1794·074	1794·444	1794·815	1795·185	1795·556	1795·926	484
485	1796·296	1796·667	1797·037	1797·407	1797·778	1798·148	1798·519	1798·889	1799·259	1799·630	485
486	1800·	1800·370	1800·741	1801·111	1801·481	1801·852	1802·222	1802·593	1802·963	1803·333	486
487	1803·704	1804·074	1804·444	1804·815	1805·185	1805·556	1805·926	1806·296	1806·667	1807·037	487
488	1807·407	1807·778	1808·148	1808·519	1808·889	1809·259	1809·630	1810·	1810·370	1810·741	488
489	1811·111	1811·481	1811·852	1812·222	1812·593	1812·963	1813·333	1813·704	1814·074	1814·444	489
490	1814·815	1815·185	1815·556	1815·926	1816·296	1816·667	1817·037	1817·407	1817·778	1818·148	490
491	1818·519	1818·889	1819·259	1819·630	1820·	1820·370	1820·741	1821·111	1821·481	1821·852	491
492	1822·222	1822·593	1822·963	1823·333	1823·704	1824·074	1824·444	1824·815	1825·185	1825·556	492
493	1825·926	1826·296	1826·667	1827·037	1827·407	1827·778	1828·148	1828·519	1828·889	1829·259	493
494	1829·630	1830·	1830·370	1830·741	1831·111	1831·481	1831·852	1832·222	1832·593	1832·963	494
495	1833·333	1833·704	1834·074	1834·444	1834·815	1835·185	1835·556	1835·926	1836·296	1836·667	495
496	1837·037	1837·407	1837·778	1838·148	1838·519	1838·889	1839·259	1839·630	1840·	1840·370	496
497	1840·741	1841·111	1841·481	1841·852	1842·222	1842·593	1842·963	1843·333	1843·704	1844·074	497
498	1844·444	1844·815	1845·185	1845·556	1845·926	1846·296	1846·667	1847·037	1847·407	1847·778	498
499	1848·148	1848·519	1848·889	1849·259	1849·630	1850·	1850·370	1850·741	1851·111	1851·481	499
500	1851·852	1852·222	1852·593	1852·963	1853·333	1853·704	1854·074	1854·444	1854·815	1855·185	500
501	1855·556	1855·926	1856·296	1856·667	1857·037	1857·407	1857·778	1858·148	1858·519	1858·889	501
502	1859·259	1859·630	1860·	1860·370	1860·741	1861·111	1861·481	1861·852	1862·222	1862·593	502
503	1862·963	1863·333	1863·704	1864·074	1864·444	1864·815	1865·185	1865·556	1865·926	1866·296	503
504	1866·667	1867·037	1867·407	1867·778	1868·148	1868·519	1868·889	1869·259	1869·630	1870·	504
505	1870·370	1870·741	1871·111	1871·481	1871·852	1872·222	1872·593	1872·963	1873·333	1873·704	505
506	1874·074	1874·444	1874·815	1875·185	1875·556	1875·926	1876·296	1876·667	1877·037	1877·407	506
507	1877·778	1878·148	1878·519	1878·889	1879·259	1879·630	1880·	1880·370	1880·741	1881·111	507
508	1881·481	1881·852	1882·222	1882·593	1882·963	1883·333	1883·704	1884·074	1884·444	1884·815	508
509	1885·185	1885·556	1885·926	1886·296	1886·667	1887·037	1887·407	1887·778	1888·148	1888·519	509
510	1888·889	1889·259	1889·630	1890·	1890·370	1890·741	1891·111	1891·481	1891·852	1892·222	510
511	1892·593	1892·963	1893·333	1893·704	1894·074	1894·444	1894·815	1895·185	1895·556	1895·926	511
512	1896·296	1896·667	1897·037	1897·407	1897·778	1898·148	1898·519	1898·889	1899·259	1899·630	512
513	1900·	1900·370	1900·741	1901·111	1901·481	1901·852	1902·222	1902·593	1902·963	1903·333	513
514	1903·704	1904·074	1904·444	1904·815	1905·185	1905·556	1905·926	1906·296	1906·667	1907·037	514
515	1907·407	1907·778	1908·148	1908·519	1908·889	1909·259	1909·630	1910·	1910·370	1910·741	515
516	1911·111	1911·481	1911·852	1912·222	1912·593	1912·963	1913·333	1913·704	1914·074	1914·444	516
517	1914·815	1915·185	1915·556	1915·926	1916·296	1916·667	1917·037	1917·407	1917·778	1918·148	517
518	1918·519	1918·889	1919·259	1919·630	1920·	1920·370	1920·741	1921·111	1921·481	1921·852	518
519	1922·222	1922·593	1922·963	1923·333	1923·704	1924·074	1924·444	1924·815	1925·185	1925·556	519
520	1925·926	1926·296	1926·667	1927·037	1927·407	1927·778	1928·148	1928·519	1928·889	1929·259	520
521	1929·630	1930·	1930·370	1930·741	1931·111	1931·481	1931·852	1932·222	1932·593	1932·963	521
522	1933·333	1933·704	1934·074	1934·444	1934·815	1935·185	1935·556	1935·926	1936·296	1936·667	522
523	1937·037	1937·407	1937·778	1938·148	1938·519	1938·889	1939·259	1939·630	1940·	1940·370	523
524	1940·741	1941·111	1941·481	1941·852	1942·222	1942·593	1942·963	1943·333	1943·704	1944·074	524
525	1944·444	1944·815	1945·185	1945·556	1945·926	1946·296	1946·667	1947·037	1947·407	1947·778	525
526	1948·148	1948·519	1948·889	1949·259	1949·630	1950·	1950·370	1950·741	1951·111	1951·481	526
527	1951·852	1952·222	1952·593	1952·963	1953·333	1953·704	1954·074	1954·444	1954·815	1955·185	527
528	1955·556	1955·926	1956·296	1956·667	1957·037	1957·407	1957·778	1958·148	1958·519	1958·889	528
529	1959·259	1959·630	1960·	1960·370	1960·741	1961·111	1961·481	1961·852	1962·222	1962·593	529
530	1962·963	1963·333	1963·704	1964·074	1964·444	1964·815	1965·185	1965·556	1965·926	1966·296	530
531	1966·667	1967·037	1967·407	1967·778	1968·148	1968·519	1968·889	1969·259	1969·630	1970·	531
532	1970·370	1970·741	1971·111	1971·481	1971·852	1972·222	1972·593	1972·963	1973·333	1973·704	532
533	1974·074	1974·444	1974·815	1975·185	1975·556	1975·926	1976·296	1976·667	1977·037	1977·407	533
534	1977·778	1978·148	1978·519	1978·889	1979·259	1979·630	1980·	1980·370	1980·741	1981·111	534
535	1981·481	1981·852	1982·222	1982·593	1982·963	1983·333	1983·704	1984·074	1984·444	1984·815	535
536	1985·185	1985·556	1985·926	1986·296	1986·667	1987·037	1987·407	1987·778	1988·148	1988·519	536
537	1988·889	1989·259	1989·630	1990·	1990·370	1990·741	1991·111	1991·481	1991·852	1992·222	537
538	1992·593	1992·963	1993·333	1993·704	1994·074	1994·444	1994·815	1995·185	1995·556	1995·926	538
539	1996·296	1996·667	1997·037	1997·407	1997·778	1998·148	1998·519	1998·889	1999·259	1999·630	539
540	2000·	2000·370	2000·741	2001·111	2001·481	2001·852	2002·222	2002·593	2002·963	2003·333	540
M.A.	·0	·1	·2	·3	·4	·5	·6	·7	·8	·9	M.A.

MEAN AREAS 481 to 540.

CUBIC YARDS TO MEAN AREAS FOR 100 FEET IN LENGTH.

M.A.	·0	·1	·2	·3	·4	·5	·6	·7	·8	·9	M.A.
541	2003·704	2004·074	2004·444	2004·815	2005·185	2005·556	2005·926	2006·296	2006·667	2007·037	541
542	2007·407	2007·778	2008·148	2008·519	2008·889	2009·259	2009·630	2010·	2010·370	2010·741	542
543	2011·111	2011·481	2011·852	2012·222	2012·593	2012·963	2013·333	2013·704	2014·074	2014·444	543
544	2014·815	2015·185	2015·556	2015·926	2016·296	2016·667	2017·037	2017·407	2017·778	2018·148	544
545	2018·519	2018·889	2019·259	2019·630	2020·	2020·370	2020·741	2021·111	2021·481	2021·852	545
546	2022·222	2022·593	2022·963	2023·333	2023·704	2024·074	2024·444	2024·815	2025·185	2025·556	546
547	2025·926	2026·296	2026·667	2027·037	2027·407	2027·778	2028·148	2028·519	2028·889	2029·259	547
548	2029·630	2030·	2030·370	2030·741	2031·111	2031·481	2031·852	2032·222	2032·593	2032·963	548
549	2033·333	2033·704	2034·074	2034·444	2034·815	2035·185	2035·556	2035·926	2036·296	2036·667	549
550	2037·037	2037·407	2037·778	2038·148	2038·519	2038·889	2039·259	2039·630	2040·	2040·370	550
551	2040·741	2041·111	2041·481	2041·852	2042·222	2042·593	2042·963	2043·333	2043·704	2044·074	551
552	2044·444	2044·815	2045·185	2045·556	2045·926	2046·296	2046·667	2047·037	2047·407	2047·778	552
553	2048·148	2048·519	2048·889	2049·259	2049·630	2050·	2050·370	2050·741	2051·111	2051·481	553
554	2051·852	2052·222	2052·593	2052·963	2053·333	2053·704	2054·074	2054·444	2054·815	2055·185	554
555	2055·556	2055·926	2056·296	2056·667	2057·037	2057·407	2057·778	2058·148	2058·519	2058·889	555
556	2059·259	2059·630	2060·	2060·370	2060·741	2061·111	2061·481	2061·852	2062·222	2062·593	556
557	2062·963	2063·333	2063·704	2064·074	2064·444	2064·815	2065·185	2065·556	2065·926	2066·296	557
558	2066·667	2067·037	2067·407	2067·778	2068·148	2068·519	2068·889	2069·259	2069·630	2070·	558
559	2070·370	2070·741	2071·111	2071·481	2071·852	2072·222	2072·593	2072·963	2073·333	2073·704	559
560	2074·074	2074·444	2074·815	2075·185	2075·556	2075·926	2076·296	2076·667	2077·037	2077·407	560
561	2077·778	2078·148	2078·519	2078·889	2079·259	2079·630	2080·	2080·370	2080·741	2081·111	561
562	2081·481	2081·852	2082·222	2082·593	2082·963	2083·333	2083·704	2084·074	2084·444	2084·815	562
563	2085·185	2085·556	2085·926	2086·296	2086·667	2087·037	2087·407	2087·778	2088·148	2088·519	563
564	2088·889	2089·259	2089·630	2090·	2090·370	2090·741	2091·111	2091·481	2091·852	2092·222	564
565	2092·593	2092·963	2093·333	2093·704	2094·074	2094·444	2094·815	2095·185	2095·556	2095·926	565
566	2096·296	2096·667	2097·037	2097·407	2097·778	2098·148	2098·519	2098·889	2099·259	2099·630	566
567	2100·	2100·370	2100·741	2101·111	2101·481	2101·852	2102·222	2102·593	2102·963	2103·333	567
568	2103·704	2104·074	2104·444	2104·815	2105·185	2105·556	2105·926	2106·296	2106·667	2107·037	568
569	2107·407	2107·778	2108·148	2108·519	2108·889	2109·259	2109·630	2110·	2110·370	2110·741	569
570	2111·111	2111·481	2111·852	2112·222	2112·593	2112·963	2113·333	2113·704	2114·074	2114·444	570
571	2114·815	2115·185	2115·556	2115·926	2116·296	2116·667	2117·037	2117·407	2117·778	2118·148	571
572	2118·519	2118·889	2119·259	2119·630	2120·	2120·370	2120·741	2121·111	2121·481	2121·852	572
573	2122·222	2122·593	2122·963	2123·333	2123·704	2124·074	2124·444	2124·815	2125·185	2125·556	573
574	2125·926	2126·296	2126·667	2127·037	2127·407	2127·778	2128·148	2128·519	2128·889	2129·259	574
575	2129·630	2130·	2130·370	2130·741	2131·111	2131·481	2131·852	2132·222	2132·593	2132·963	575
576	2133·333	2133·704	2134·074	2134·444	2134·815	2135·185	2135·556	2135·926	2136·296	2136·667	576
577	2137·037	2137·407	2137·778	2138·148	2138·519	2138·889	2139·259	2139·630	2140·	2140·370	577
578	2140·741	2141·111	2141·481	2141·852	2142·222	2142·593	2142·963	2143·333	2143·704	2144·074	578
579	2144·444	2144·815	2145·185	2145·556	2145·926	2146·296	2146·667	2147·037	2147·407	2147·778	579
580	2148·148	2148·519	2148·889	2149·259	2149·630	2150·	2150·370	2150·741	2151·111	2151·481	580
581	2151·852	2152·222	2152·593	2152·963	2153·333	2153·704	2154·074	2154·444	2154·815	2155·185	581
582	2155·556	2155·926	2156·296	2156·667	2157·037	2157·407	2157·778	2158·148	2158·519	2158·889	582
583	2159·259	2159·630	2160·	2160·370	2160·741	2161·111	2161·481	2161·852	2162·222	2162·593	583
584	2162·963	2163·333	2163·704	2164·074	2164·444	2164·815	2165·185	2165·556	2165·926	2166·296	584
585	2166·667	2167·037	2167·407	2167·778	2168·148	2168·519	2168·889	2169·259	2169·630	2170·	585
586	2170·370	2170·741	2171·111	2171·481	2171·852	2172·222	2172·593	2172·963	2173·333	2173·704	586
587	2174·074	2174·444	2174·815	2175·185	2175·556	2175·926	2176·296	2176·667	2177·037	2177·407	587
588	2177·778	2178·148	2178·519	2178·889	2179·259	2179·630	2180·	2180·370	2180·741	2181·111	588
589	2181·481	2181·852	2182·222	2182·593	2182·963	2183·333	2183·704	2184·074	2184·444	2184·815	589
590	2185·185	2185·556	2185·926	2186·296	2186·667	2187·037	2187·407	2187·778	2188·148	2188·519	590
591	2188·889	2189·259	2189·630	2190·	2190·370	2190·741	2191·111	2191·481	2191·852	2192·222	591
592	2192·593	2192·963	2193·333	2193·704	2194·074	2194·444	2194·815	2195·185	2195·556	2195·926	592
593	2196·296	2196·667	2197·037	2197·407	2197·778	2198·148	2198·519	2198·889	2199·259	2199·630	593
594	2200·	2200·370	2200·741	2201·111	2201·481	2201·852	2202·222	2202·593	2202·963	2203·333	594
595	2203·704	2204·074	2204·444	2204·815	2205·185	2205·556	2205·926	2206·296	2206·667	2207·037	595
596	2207·407	2207·778	2208·148	2208·519	2208·889	2209·259	2209·630	2210·	2210·370	2210·741	596
597	2211·111	2211·481	2211·852	2212·222	2212·593	2212·963	2213·333	2213·704	2214·074	2214·444	597
598	2214·815	2215·185	2215·556	2215·926	2216·296	2216·667	2217·037	2217·407	2217·778	2218·148	598
599	2218·519	2218·889	2219·259	2219·630	2220·	2220·370	2220·741	2221·111	2221·481	2221·852	599
600	2222·222	2222·593	2222·963	2223·333	2223·704	2224·074	2224·444	2224·815	2225·185	2225·556	600
M.A.	·0	·1	·2	·3	·4	·5	·6	·7	·8	·9	M.A.

CUBIC YARDS TO MEAN AREAS FOR 100 FEET IN LENGTH.

M.A.	·0	·1	·2	·3	·4	·5	·6	·7	·8	·9	M.A.
601											601
602											602
603											603
604											604
605											605
606											606
607											607
608											608
609											609
610											610
611											611
612											612
613											613
614											614
615											615
616											616
617											617
618											618
619											619
620											620
621											621
622											622
623											623
624											624
625											625
626											626
627											627
628											628
629											629
630											630
631											631
632											632
633											633
634											634
635											635
636											636
637											637
638											638
639											639
640											640
641											641
642											642
643											643
644											644
645											645
646											646
647											647
648											648
649											649
650											650
651											651
652											652
653											653
654											654
655											655
656											656
657											657
658											658
659											659
660											660
M.A.	·0	·1	·2	·3	·4	·5	·6	·7	·8	·9	M.A.

MEAN AREAS 601 to 660.

CUBIC YARDS TO MEAN AREAS FOR 100 FEET IN LENGTH.

M.A.	·0	·1	·2	·3	·4	·5	·6	·7	·8	·9	M.A.
661	2448·148	2448·519	2448·889	2449·259	2449·630	2450·	2450·370	2450·741	2451·111	2451·481	661
662	2451·852	2452·222	2452·593	2452·963	2453·333	2453·704	2454·074	2454·444	2454·815	2455·185	662
663	2455·556	2455·926	2456·296	2456·667	2457·037	2457·407	2457·778	2458·148	2458·519	2458·889	663
664	2459·259	2459·630	2460·	2460·370	2460·741	2461·111	2461·481	2461·852	2462·222	2462·593	664
665	2462·963	2463·333	2463·704	2464·074	2464·444	2464·815	2465·185	2465·556	2465·926	2466·296	665
666	2466·667	2467·037	2467·407	2467·778	2468·148	2468·519	2468·889	2469·259	2469·630	2470·	666
667	2470·370	2470·741	2471·111	2471·481	2471·852	2472·222	2472·593	2472·963	2473·333	2473·704	667
668	2474·074	2474·444	2474·815	2475·185	2475·556	2475·926	2476·296	2476·667	2477·037	2477·407	668
669	2477·778	2478·148	2478·519	2478·889	2479·259	2479·630	2480·	2480·370	2480·741	2481·111	669
670	2481·481	2481·852	2482·222	2482·593	2482·963	2483·333	2483·704	2484·074	2484·444	2484·815	670
671	2485·185	2485·556	2485·926	2486·296	2486·667	2487·037	2487·407	2487·778	2488·148	2488·519	671
672	2488·889	2489·259	2489·630	2490·	2490·370	2490·741	2491·111	2491·481	2491·852	2492·222	672
673	2492·593	2492·963	2493·333	2493·704	2494·074	2494·444	2494·815	2495·185	2495·556	2495·926	673
674	2496·296	2496·667	2497·037	2497·407	2497·778	2498·148	2498·519	2498·889	2499·259	2499·630	674
675	2500·	2500·370	2500·741	2501·111	2501·481	2501·852	2502·222	2502·593	2502·963	2503·333	675
676	2503·704	2504·074	2504·444	2504·815	2505·185	2505·556	2505·926	2506·296	2506·667	2507·037	676
677	2507·407	2507·778	2508·148	2508·519	2508·889	2509·259	2509·630	2510·	2510·370	2510·741	677
678	2511·111	2511·481	2511·852	2512·222	2512·593	2512·963	2513·333	2513·704	2514·074	2514·444	678
679	2514·815	2515·185	2515·556	2515·926	2516·296	2516·667	2517·037	2517·407	2517·778	2518·148	679
680	2518·519	2518·889	2519·259	2519·630	2520·	2520·370	2520·741	2521·111	2521·481	2521·852	680
681	2522·222	2522·593	2522·963	2523·333	2523·704	2524·074	2524·444	2524·815	2525·185	2525·556	681
682	2525·926	2526·296	2526·667	2527·037	2527·407	2527·778	2528·148	2528·519	2528·889	2529·259	682
683	2529·630	2530·	2530·370	2530·741	2531·111	2531·481	2531·852	2532·222	2532·593	2532·963	683
684	2533·333	2533·704	2534·074	2534·444	2534·815	2535·185	2535·556	2535·926	2536·296	2536·667	684
685	2537·037	2537·407	2537·778	2538·148	2538·519	2538·889	2539·259	2539·630	2540·	2540·370	685
686	2540·741	2541·111	2541·481	2541·852	2542·222	2542·593	2542·963	2543·333	2543·704	2544·074	686
687	2544·444	2544·815	2545·185	2545·556	2545·926	2546·296	2546·667	2547·037	2547·407	2547·778	687
688	2548·148	2548·519	2548·889	2549·259	2549·630	2550·	2550·370	2550·741	2551·111	2551·481	688
689	2551·852	2552·222	2552·593	2552·963	2553·333	2553·704	2554·074	2554·444	2554·815	2555·185	689
690	2555·556	2555·926	2556·296	2556·667	2557·037	2557·407	2557·778	2558·148	2558·519	2558·889	690
691	2559·259	2559·630	2560·	2560·370	2560·741	2561·111	2561·481	2561·852	2562·222	2562·593	691
692	2562·963	2563·333	2563·704	2564·074	2564·444	2564·815	2565·185	2565·556	2565·926	2566·296	692
693	2566·667	2567·037	2567·407	2567·778	2568·148	2568·519	2568·889	2569·259	2569·630	2570·	693
694	2570·370	2570·741	2571·111	2571·481	2571·852	2572·222	2572·593	2572·963	2573·333	2573·704	694
695	2574·074	2574·444	2574·815	2575·185	2575·556	2575·926	2576·296	2576·667	2577·037	2577·407	695
696	2577·778	2578·148	2578·519	2578·889	2579·259	2579·630	2580·	2580·370	2580·741	2581·111	696
697	2581·481	2581·852	2582·222	2582·593	2582·963	2583·333	2583·704	2584·074	2584·444	2584·815	697
698	2585·185	2585·556	2585·926	2586·296	2586·667	2587·037	2587·407	2587·778	2588·148	2588·519	698
699	2588·889	2589·259	2589·630	2590·	2590·370	2590·741	2591·111	2591·481	2591·852	2592·222	699
700	2592·593	2592·963	2593·333	2593·704	2594·074	2594·444	2594·815	2595·185	2595·556	2595·926	700
701	2596·296	2596·667	2597·037	2597·407	2597·778	2598·148	2598·519	2598·889	2599·259	2599·630	701
702	2600·	2600·370	2600·741	2601·111	2601·481	2601·852	2602·222	2602·593	2602·963	2603·333	702
703	2603·704	2604·074	2604·444	2604·815	2605·185	2605·556	2605·926	2606·296	2606·667	2607·037	703
704	2607·407	2607·778	2608·148	2608·519	2608·889	2609·259	2609·630	2610·	2610·370	2610·741	704
705	2611·111	2611·481	2611·852	2612·222	2612·593	2612·963	2613·333	2613·704	2614·074	2614·444	705
706	2614·815	2615·185	2615·556	2615·926	2616·296	2616·667	2617·037	2617·407	2617·778	2618·148	706
707	2618·519	2618·889	2619·259	2619·630	2620·	2620·370	2620·741	2621·111	2621·481	2621·852	707
708	2622·222	2622·593	2622·963	2623·333	2623·704	2624·074	2624·444	2624·815	2625·185	2625·556	708
709	2625·926	2626·296	2626·667	2627·037	2627·407	2627·778	2628·148	2628·519	2628·889	2629·259	709
710	2629·630	2630·	2630·370	2630·741	2631·111	2631·481	2631·852	2632·222	2632·593	2632·963	710
711	2633·333	2633·704	2634·074	2634·444	2634·815	2635·185	2635·556	2635·926	2636·296	2636·667	711
712	2637·037	2637·407	2637·778	2638·148	2638·519	2638·889	2639·259	2639·630	2640·	2640·370	712
713	2640·741	2641·111	2641·481	2641·852	2642·222	2642·593	2642·963	2643·333	2643·704	2644·074	713
714	2644·444	2644·815	2645·185	2645·556	2645·926	2646·296	2646·667	2647·037	2647·407	2647·778	714
715	2648·148	2648·519	2648·889	2649·259	2649·630	2650·	2650·370	2650·741	2651·111	2651·481	715
716	2651·852	2652·222	2652·593	2652·963	2653·333	2653·704	2654·074	2654·444	2654·815	2655·185	716
717	2655·556	2655·926	2656·296	2656·667	2657·037	2657·407	2657·778	2658·148	2658·519	2658·889	717
718	2659·259	2659·630	2660·	2660·370	2660·741	2661·111	2661·481	2661·852	2662·222	2662·593	718
719	2662·963	2663·333	2663·704	2664·074	2664·444	2664·815	2665·185	2665·556	2665·926	2666·296	719
720	2666·667	2667·037	2667·407	2667·778	2668·148	2668·519	2668·889	2669·259	2669·630	2670·	720
M.A.	·0	·1	·2	·3	·4	·5	·6	·7	·8	·9	M.A.

MEAN AREAS 661 to 720.

CUBIC YARDS TO MEAN AREAS FOR 100 FEET IN LENGTH.

M.A.	·0	·1	·2	·3	·4	·5	·6	·7	·8	·9	M.A.
721	2670·370	2670·741	2671·111	2671·481	2671·852	2672·222	2672·593	2672·963	2673·333	2673·704	721
722	2674·074	2674·444	2674·815	2675·185	2675·556	2675·926	2676·296	2676·667	2677·037	2677·407	722
723	2677·778	2678·148	2678·519	2678·889	2679·259	2679·630	2680·000	2680·370	2680·741	2681·111	723
724	2681·481	2681·852	2682·222	2682·593	2682·963	2683·333	2683·704	2684·074	2684·444	2684·815	724
725	2685·185	2685·556	2685·926	2686·296	2686·667	2687·037	2687·407	2687·778	2688·148	2688·519	725
726	2688·889	2689·259	2689·630	2690·000	2690·370	2690·741	2691·111	2691·481	2691·852	2692·222	726
727	2692·593	2692·963	2693·333	2693·704	2694·074	2694·444	2694·815	2695·185	2695·556	2695·926	727
728	2696·296	2696·667	2697·037	2697·407	2697·778	2698·148	2698·519	2698·889	2699·259	2699·630	728
729	2700·000	2700·370	2700·741	2701·111	2701·481	2701·852	2702·222	2702·593	2702·963	2703·333	729
730	2703·704	2704·074	2704·444	2704·815	2705·185	2705·556	2705·926	2706·296	2706·667	2707·037	730
731	2707·407	2707·778	2708·148	2708·519	2708·889	2709·259	2709·630	2710·000	2710·370	2710·741	731
732	2711·111	2711·481	2711·852	2712·222	2712·593	2712·963	2713·333	2713·704	2714·074	2714·444	732
733	2714·815	2715·185	2715·556	2715·926	2716·296	2716·667	2717·037	2717·407	2717·778	2718·148	733
734	2718·519	2718·889	2719·259	2719·630	2720·000	2720·370	2720·741	2721·111	2721·481	2721·852	734
735	2722·222	2722·593	2722·963	2723·333	2723·704	2724·074	2724·444	2724·815	2725·185	2725·556	735
736	2725·926	2726·296	2726·667	2727·037	2727·407	2727·778	2728·148	2728·519	2728·889	2729·259	736
737	2729·630	2730·000	2730·370	2730·741	2731·111	2731·481	2731·852	2732·222	2732·593	2732·963	737
738	2733·333	2733·704	2734·074	2734·444	2734·815	2735·185	2735·556	2735·926	2736·296	2736·667	738
739	2737·037	2737·407	2737·778	2738·148	2738·519	2738·889	2739·259	2739·630	2740·000	2740·370	739
740	2740·741	2741·111	2741·481	2741·852	2742·222	2742·593	2742·963	2743·333	2743·704	2744·074	740
741	2744·444	2744·815	2745·185	2745·556	2745·926	2746·296	2746·667	2747·037	2747·407	2747·778	741
742	2748·148	2748·519	2748·889	2749·259	2749·630	2750·000	2750·370	2750·741	2751·111	2751·481	742
743	2751·852	2752·222	2752·593	2752·963	2753·333	2753·704	2754·074	2754·444	2754·815	2755·185	743
744	2755·556	2755·926	2756·296	2756·667	2757·037	2757·407	2757·778	2758·148	2758·519	2758·889	744
745	2759·259	2759·630	2760·000	2760·370	2760·741	2761·111	2761·481	2761·852	2762·222	2762·593	745
746	2762·963	2763·333	2763·704	2764·074	2764·444	2764·815	2765·185	2765·556	2765·926	2766·296	746
747	2766·667	2767·037	2767·407	2767·778	2768·148	2768·519	2768·889	2769·259	2769·630	2770·000	747
748	2770·370	2770·741	2771·111	2771·481	2771·852	2772·222	2772·593	2772·963	2773·333	2773·704	748
749	2774·074	2774·444	2774·815	2775·185	2775·556	2775·926	2776·296	2776·667	2777·037	2777·407	749
750	2777·778	2778·148	2778·519	2778·889	2779·259	2779·630	2780·000	2780·370	2780·741	2781·111	750
751	2781·481	2781·852	2782·222	2782·593	2782·963	2783·333	2783·704	2784·074	2784·444	2784·815	751
752	2785·185	2785·556	2785·926	2786·296	2786·667	2787·037	2787·407	2787·778	2788·148	2788·519	752
753	2788·889	2789·259	2789·630	2790·000	2790·370	2790·741	2791·111	2791·481	2791·852	2792·222	753
754	2792·593	2792·963	2793·333	2793·704	2794·074	2794·444	2794·815	2795·185	2795·556	2795·926	754
755	2796·296	2796·667	2797·037	2797·407	2797·778	2798·148	2798·519	2798·889	2799·259	2799·630	755
756	2800·000	2800·370	2800·741	2801·111	2801·481	2801·852	2802·222	2802·593	2802·963	2803·333	756
757	2803·704	2804·074	2804·444	2804·815	2805·185	2805·556	2805·926	2806·296	2806·667	2807·037	757
758	2807·407	2807·778	2808·148	2808·519	2808·889	2809·259	2809·630	2810·000	2810·370	2810·741	758
759	2811·111	2811·481	2811·852	2812·222	2812·593	2812·963	2813·333	2813·704	2814·074	2814·444	759
760	2814·815	2815·185	2815·556	2815·926	2816·296	2816·667	2817·037	2817·407	2817·778	2818·148	760
761	2818·519	2818·889	2819·259	2819·630	2820·000	2820·370	2820·741	2821·111	2821·481	2821·852	761
762	2822·222	2822·593	2822·963	2823·333	2823·704	2824·074	2824·444	2824·815	2825·185	2825·556	762
763	2825·926	2826·296	2826·667	2827·037	2827·407	2827·778	2828·148	2828·519	2828·889	2829·259	763
764	2829·630	2830·000	2830·370	2830·741	2831·111	2831·481	2831·852	2832·222	2832·593	2832·963	764
765	2833·333	2833·704	2834·074	2834·444	2834·815	2835·185	2835·556	2835·926	2836·296	2836·667	765
766	2837·037	2837·407	2837·778	2838·148	2838·519	2838·889	2839·259	2839·630	2840·000	2840·370	766
767	2840·741	2841·111	2841·481	2841·852	2842·222	2842·593	2842·963	2843·333	2843·704	2844·074	767
768	2844·444	2844·815	2845·185	2845·556	2845·926	2846·296	2846·667	2847·037	2847·407	2847·778	768
769	2848·148	2848·519	2848·889	2849·259	2849·630	2850·000	2850·370	2850·741	2851·111	2851·481	769
770	2851·852	2852·222	2852·593	2852·963	2853·333	2853·704	2854·074	2854·444	2854·815	2855·185	770
771	2855·556	2855·926	2856·296	2856·667	2857·037	2857·407	2857·778	2858·148	2858·519	2858·889	771
772	2859·259	2859·630	2860·000	2860·370	2860·741	2861·111	2861·481	2861·852	2862·222	2862·593	772
773	2862·963	2863·333	2863·704	2864·074	2864·444	2864·815	2865·185	2865·556	2865·926	2866·296	773
774	2866·667	2867·037	2867·407	2867·778	2868·148	2868·519	2868·889	2869·259	2869·630	2870·000	774
775	2870·370	2870·741	2871·111	2871·481	2871·852	2872·222	2872·593	2872·963	2873·333	2873·704	775
776	2874·074	2874·444	2874·815	2875·185	2875·556	2875·926	2876·296	2876·667	2877·037	2877·407	776
777	2877·778	2878·148	2878·519	2878·889	2879·259	2879·630	2880·000	2880·370	2880·741	2881·111	777
778	2881·481	2881·852	2882·222	2882·593	2882·963	2883·333	2883·704	2884·074	2884·444	2884·815	778
779	2885·185	2885·556	2885·926	2886·296	2886·667	2887·037	2887·407	2887·778	2888·148	2888·519	779
780	2888·889	2889·259	2889·630	2890·000	2890·370	2890·741	2891·111	2891·481	2891·852	2892·222	780
M.A.	·0	·1	·2	·3	·4	·5	·6	·7	·8	·9	M.A.

MEAN AREAS 721 to 780.

CUBIC YARDS TO MEAN AREAS FOR 100 FEET IN LENGTH.

M.A.	·0	·1	·2	·3	·4	·5	·6	·7	·8	·9	M.A.
781	2892·593	2892·963	2893·333	2893·704	2894·074	2894·444	2894·815	2895·185	2895·556	2895·926	781
782	2896·296	2896·667	2897·037	2897·407	2897·778	2898·148	2898·519	2898·889	2899·259	2899·630	782
783	2900·000	2900·370	2900·741	2901·111	2901·481	2901·852	2902·222	2902·593	2902·963	2903·333	783
784	2903·704	2904·074	2904·444	2904·815	2905·185	2905·556	2905·926	2906·296	2906·667	2907·037	784
785	2907·407	2907·778	2908·148	2908·519	2908·889	2909·259	2909·630	2910·000	2910·370	2910·741	785
786	2911·111	2911·481	2911·852	2912·222	2912·593	2912·963	2913·333	2913·704	2914·074	2914·444	786
787	2914·815	2915·185	2915·556	2915·926	2916·296	2916·667	2917·037	2917·407	2917·778	2918·148	787
788	2918·519	2918·889	2919·259	2919·630	2920·000	2920·370	2920·741	2921·111	2921·481	2921·852	788
789	2922·222	2922·593	2922·963	2923·333	2923·704	2924·074	2924·444	2924·815	2925·185	2925·556	789
790	2925·926	2926·296	2926·667	2927·037	2927·407	2927·778	2928·148	2928·519	2928·889	2929·259	790
791	2929·630	2930·000	2930·370	2930·741	2931·111	2931·481	2931·852	2932·222	2932·593	2932·963	791
792	2933·333	2933·704	2934·074	2934·444	2934·815	2935·185	2935·556	2935·926	2936·296	2936·667	792
793	2937·037	2937·407	2937·778	2938·148	2938·519	2938·889	2939·259	2939·630	2940·000	2940·370	793
794	2940·741	2941·111	2941·481	2941·852	2942·222	2942·593	2942·963	2943·333	2943·704	2944·074	794
795	2944·444	2944·815	2945·185	2945·556	2945·926	2946·296	2946·667	2947·037	2947·407	2947·778	795
796	2948·148	2948·519	2948·889	2949·259	2949·630	2950·000	2950·370	2950·741	2951·111	2951·481	796
797	2951·852	2952·222	2952·593	2952·963	2953·333	2953·704	2954·074	2954·444	2954·815	2955·185	797
798	2955·556	2955·926	2956·296	2956·667	2957·037	2957·407	2957·778	2958·148	2958·519	2958·889	798
799	2959·259	2959·630	2960·000	2960·370	2960·741	2961·111	2961·481	2961·852	2962·222	2962·593	799
800	2962·963	2963·333	2963·704	2964·074	2964·444	2964·815	2965·185	2965·556	2965·926	2966·296	800
801	2966·667	2967·037	2967·407	2967·778	2968·148	2968·519	2968·889	2969·259	2969·630	2970·000	801
802	2970·370	2970·741	2971·111	2971·481	2971·852	2972·222	2972·593	2972·963	2973·333	2973·704	802
803	2974·074	2974·444	2974·815	2975·185	2975·556	2975·926	2976·296	2976·667	2977·037	2977·407	803
804	2977·778	2978·148	2978·519	2978·889	2979·259	2979·630	2980·000	2980·370	2980·741	2981·111	804
805	2981·481	2981·852	2982·222	2982·593	2982·963	2983·333	2983·704	2984·074	2984·444	2984·815	805
806	2985·185	2985·556	2985·926	2986·296	2986·667	2987·037	2987·407	2987·778	2988·148	2988·519	806
807	2988·889	2989·259	2989·630	2990·000	2990·370	2990·741	2991·111	2991·481	2991·852	2992·222	807
808	2992·593	2992·963	2993·333	2993·704	2994·074	2994·444	2994·815	2995·185	2995·556	2995·926	808
809	2996·296	2996·667	2997·037	2997·407	2997·778	2998·148	2998·519	2998·889	2999·259	2999·630	809
810	3000·000	3000·370	3000·741	3001·111	3001·481	3001·852	3002·222	3002·593	3002·963	3003·333	810
811	3003·704	3004·074	3004·444	3004·815	3005·185	3005·556	3005·926	3006·296	3006·667	3007·037	811
812	3007·407	3007·778	3008·148	3008·519	3008·889	3009·259	3009·630	3010·000	3010·370	3010·741	812
813	3011·111	3011·481	3011·852	3012·222	3012·593	3012·963	3013·333	3013·704	3014·074	3014·444	813
814	3014·815	3015·185	3015·556	3015·926	3016·296	3016·667	3017·037	3017·407	3017·778	3018·148	814
815	3018·519	3018·889	3019·259	3019·630	3020·000	3020·370	3020·741	3021·111	3021·481	3021·852	815
816	3022·222	3022·593	3022·963	3023·333	3023·704	3024·074	3024·444	3024·815	3025·185	3025·556	816
817	3025·926	3026·296	3026·667	3027·037	3027·407	3027·778	3028·148	3028·519	3028·889	3029·259	817
818	3029·630	3030·000	3030·370	3030·741	3031·111	3031·481	3031·852	3032·222	3032·593	3032·963	818
819	3033·333	3033·704	3034·074	3034·444	3034·815	3035·185	3035·556	3035·926	3036·296	3036·667	819
820	3037·037	3037·407	3037·778	3038·148	3038·519	3038·889	3039·259	3039·630	3040·000	3040·370	820
821	3040·741	3041·111	3041·481	3041·852	3042·222	3042·593	3042·963	3043·333	3043·704	3044·074	821
822	3044·444	3044·815	3045·185	3045·556	3045·926	3046·296	3046·667	3047·037	3047·407	3047·778	822
823	3048·148	3048·519	3048·889	3049·259	3049·630	3050·000	3050·370	3050·741	3051·111	3051·481	823
824	3051·852	3052·222	3052·593	3052·963	3053·333	3053·704	3054·074	3054·444	3054·815	3055·185	824
825	3055·556	3055·926	3056·296	3056·667	3057·037	3057·407	3057·778	3058·148	3058·519	3058·889	825
826	3059·259	3059·630	3060·000	3060·370	3060·741	3061·111	3061·481	3061·852	3062·222	3062·593	826
827	3062·963	3063·333	3063·704	3064·074	3064·444	3064·815	3065·185	3065·556	3065·926	3066·296	827
828	3066·667	3067·037	3067·407	3067·778	3068·148	3068·519	3068·889	3069·259	3069·630	3070·000	828
829	3070·370	3070·741	3071·111	3071·481	3071·852	3072·222	3072·593	3072·963	3073·333	3073·704	829
830	3074·074	3074·444	3074·815	3075·185	3075·556	3075·926	3076·296	3076·667	3077·037	3077·407	830
831	3077·778	3078·148	3078·519	3078·889	3079·259	3079·630	3080·000	3080·370	3080·741	3081·111	831
832	3081·481	3081·852	3082·222	3082·593	3082·963	3083·333	3083·704	3084·074	3084·444	3084·815	832
833	3085·185	3085·556	3085·926	3086·296	3086·667	3087·037	3087·407	3087·778	3088·148	3088·519	833
834	3088·889	3089·259	3089·630	3090·000	3090·370	3090·741	3091·111	3091·481	3091·852	3092·222	834
835	3092·593	3092·963	3093·333	3093·704	3094·074	3094·444	3094·815	3095·185	3095·556	3095·926	835
836	3096·296	3096·667	3097·037	3097·407	3097·778	3098·148	3098·519	3098·889	3099·259	3099·630	836
837	3100·000	3100·370	3100·741	3101·111	3101·481	3101·852	3102·222	3102·593	3102·963	3103·333	837
838	3103·704	3104·074	3104·444	3104·815	3105·185	3105·556	3105·926	3106·296	3106·667	3107·037	838
839	3107·407	3107·778	3108·148	3108·519	3108·889	3109·259	3109·630	3110·000	3110·370	3110·741	839
840	3111·111	3111·481	3111·852	3112·222	3112·593	3112·963	3113·333	3113·704	3114·074	3114·444	840
M.A.	·0	·1	·2	·3	·4	·5	·6	·7	·8	·9	M.A.

MEAN AREAS 781 to 840.

CUBIC YARDS TO MEAN AREAS FOR 100 FEET IN LENGTH.

M.A.	·0	·1	·2	·3	·4	·5	·6	·7	·8	·9	M.A.
841	3114·815	3115·185	3115·556	3115·926	3116·296	3116·667	3117·037	3117·407	3117·778	3118·148	841
842	3118·519	3118·889	3119·259	3119·630	3120·	3120·370	3120·741	3121·111	3121·481	3121·852	842
843	3122·222	3122·593	3122·963	3123·333	3123·704	3124·074	3124·444	3124·815	3125·185	3125·556	843
844	3125·926	3126·296	3126·667	3127·037	3127·407	3127·778	3128·148	3128·519	3128·889	3129·259	844
845	3129·630	3130·	3130·370	3130·741	3131·111	3131·481	3131·852	3132·222	3132·593	3132·963	845
846	3133·333	3133·704	3134·074	3134·444	3134·815	3135·185	3135·556	3135·926	3136·296	3136·667	846
847	3137·037	3137·407	3137·778	3138·148	3138·519	3138·889	3139·259	3139·630	3140·	3140·370	847
848	3140·741	3141·111	3141·481	3141·852	3142·222	3142·593	3142·963	3143·333	3143·704	3144·074	848
849	3144·444	3144·815	3145·185	3145·556	3145·926	3146·296	3146·667	3147·037	3147·407	3147·778	849
850	3148·148	3148·519	3148·889	3149·259	3149·630	3150·	3150·370	3150·741	3151·111	3151·481	850
851	3151·852	3152·222	3152·593	3152·963	3153·333	3153·704	3154·074	3154·444	3154·815	3155·185	851
852	3155·556	3155·926	3156·296	3156·667	3157·037	3157·407	3157·778	3158·148	3158·519	3158·889	852
853	3159·259	3159·630	3160·	3160·370	3160·741	3161·111	3161·481	3161·852	3162·222	3162·593	853
854	3162·963	3163·333	3163·704	3164·074	3164·444	3164·815	3165·185	3165·556	3165·926	3166·296	854
855	3166·667	3167·037	3167·407	3167·778	3168·148	3168·519	3168·889	3169·259	3169·630	3170·	855
856	3170·370	3170·741	3171·111	3171·481	3171·852	3172·222	3172·593	3172·963	3173·333	3173·704	856
857	3174·074	3174·444	3174·815	3175·185	3175·556	3175·926	3176·296	3176·667	3177·037	3177·407	857
858	3177·778	3178·148	3178·519	3178·889	3179·259	3179·630	3180·	3180·370	3180·741	3181·111	858
859	3181·481	3181·852	3182·222	3182·593	3182·963	3183·333	3183·704	3184·074	3184·444	3184·815	859
860	3185·185	3185·556	3185·926	3186·296	3186·667	3187·037	3187·407	3187·778	3188·148	3188·519	860
861	3188·889	3189·259	3189·630	3190·	3190·370	3190·741	3191·111	3191·481	3191·852	3192·222	861
862	3192·593	3192·963	3193·333	3193·704	3194·074	3194·444	3194·815	3195·185	3195·556	3195·926	862
863	3196·296	3196·667	3197·037	3197·407	3197·778	3198·148	3198·519	3198·889	3199·259	3199·630	863
864	3200·	3200·370	3200·741	3201·111	3201·481	3201·852	3202·222	3202·593	3202·963	3203·333	864
865	3203·704	3204·074	3204·444	3204·815	3205·185	3205·556	3205·926	3206·296	3206·667	3207·037	865
866	3207·407	3207·778	3208·148	3208·519	3208·889	3209·259	3209·630	3210·	3210·370	3210·741	866
867	3211·111	3211·481	3211·852	3212·222	3212·593	3212·963	3213·333	3213·704	3214·074	3214·444	867
868	3214·815	3215·185	3215·556	3215·926	3216·296	3216·667	3217·037	3217·407	3217·778	3218·148	868
869	3218·519	3218·889	3219·259	3219·630	3220·	3220·370	3220·741	3221·111	3221·481	3221·852	869
870	3222·222	3222·593	3222·963	3223·333	3223·704	3224·074	3224·444	3224·815	3225·185	3225·556	870
871	3225·926	3226·296	3226·667	3227·037	3227·407	3227·778	3228·148	3228·519	3228·889	3229·259	871
872	3229·630	3230·	3230·370	3230·741	3231·111	3231·481	3231·852	3232·222	3232·593	3232·963	872
873	3233·333	3233·704	3234·074	3234·444	3234·815	3235·185	3235·556	3235·926	3236·296	3236·667	873
874	3237·037	3237·407	3237·778	3238·148	3238·519	3238·889	3239·259	3239·630	3240·	3240·370	874
875	3240·741	3241·111	3241·481	3241·852	3242·222	3242·593	3242·963	3243·333	3243·704	3244·074	875
876	3244·444	3244·815	3245·185	3245·556	3245·926	3246·296	3246·667	3247·037	3247·407	3247·778	876
877	3248·148	3248·519	3248·889	3249·259	3249·630	3250·	3250·370	3250·741	3251·111	3251·481	877
878	3251·852	3252·222	3252·593	3252·963	3253·333	3253·704	3254·074	3254·444	3254·815	3255·185	878
879	3255·556	3255·926	3256·296	3256·667	3257·037	3257·407	3257·778	3258·148	3258·519	3258·889	879
880	3259·259	3259·630	3260·	3260·370	3260·741	3261·111	3261·481	3261·852	3262·222	3262·593	880
881	3262·963	3263·333	3263·704	3264·074	3264·444	3264·815	3265·185	3265·556	3265·926	3266·296	881
882	3266·667	3267·037	3267·407	3267·778	3268·148	3268·519	3268·889	3269·259	3269·630	3270·	882
883	3270·370	3270·741	3271·111	3271·481	3271·852	3272·222	3272·593	3272·963	3273·333	3273·704	883
884	3274·074	3274·444	3274·815	3275·185	3275·556	3275·926	3276·296	3276·667	3277·037	3277·407	884
885	3277·778	3278·148	3278·519	3278·889	3279·259	3279·630	3280·	3280·370	3280·741	3281·111	885
886	3281·481	3281·852	3282·222	3282·593	3282·963	3283·333	3283·704	3284·074	3284·444	3284·815	886
887	3285·185	3285·556	3285·926	3286·296	3286·667	3287·037	3287·407	3287·778	3288·148	3288·519	887
888	3288·889	3289·259	3289·630	3290·	3290·370	3290·741	3291·111	3291·481	3291·852	3292·222	888
889	3292·593	3292·963	3293·333	3293·704	3294·074	3294·444	3294·815	3295·185	3295·556	3295·926	889
890	3296·296	3296·667	3297·037	3297·407	3297·778	3298·148	3298·519	3298·889	3299·259	3299·630	890
891	3300·	3300·370	3300·741	3301·111	3301·481	3301·852	3302·222	3302·593	3302·963	3303·333	891
892	3303·704	3304·074	3304·444	3304·815	3305·185	3305·556	3305·926	3306·296	3306·667	3307·037	892
893	3307·407	3307·778	3308·148	3308·519	3308·889	3309·259	3309·630	3310·	3310·370	3310·741	893
894	3311·111	3311·481	3311·852	3312·222	3312·593	3312·963	3313·333	3313·704	3314·074	3314·444	894
895	3314·815	3315·185	3315·556	3315·926	3316·296	3316·667	3317·037	3317·407	3317·778	3318·148	895
896	3318·519	3318·889	3319·259	3319·630	3320·	3320·370	3320·741	3321·111	3321·481	3321·852	896
897	3322·222	3322·593	3322·963	3323·333	3323·704	3324·074	3324·444	3324·815	3325·185	3325·556	897
898	3325·926	3326·296	3326·667	3327·037	3327·407	3327·778	3328·148	3328·519	3328·889	3329·259	898
899	3329·630	3330·	3330·370	3330·741	3331·111	3331·481	3331·852	3332·222	3332·593	3332·963	899
900	3333·333	3333·704	3334·074	3334·444	3334·815	3335·185	3335·556	3335·926	3336·296	3336·667	900
M.A.	·0	·1	·2	·3	·4	·5	·6	·7	·8	·9	M.A.

MEAN AREAS 841 to 900.

CUBIC YARDS TO MEAN AREAS FOR 100 FEET IN LENGTH.

M.A.	·0	·1	·2	·3	·4	·5	·6	·7	·8	·9	M.A.
901	3337·037	3337·407	3337·778	3338·148	3338·519	3338·889	3339·259	3339·630	3340·	3340·370	901
902	3340·741	3341·111	3341·481	3341·852	3342·222	3342·593	3342·963	3343·333	3343·704	3344·074	902
903	3344·444	3344·815	3345·185	3345·556	3345·926	3346·296	3346·667	3347·037	3347·407	3347·778	903
904	3348·148	3348·519	3348·889	3349·259	3349·630	3350·	3350·370	3350·741	3351·111	3351·481	904
905	3351·852	3352·222	3352·593	3352·963	3353·333	3353·704	3354·074	3354·444	3354·815	3355·185	905
906	3355·556	3355·926	3356·296	3356·667	3357·037	3357·407	3357·778	3358·148	3358·519	3358·889	906
907	3359·259	3359·630	3360·	3360·370	3360·741	3361·111	3361·481	3361·852	3362·222	3362·593	907
908	3362·963	3363·333	3363·704	3364·074	3364·444	3364·815	3365·185	3365·556	3365·926	3366·296	908
909	3366·667	3367·037	3367·407	3367·778	3368·148	3368·519	3368·889	3369·259	3369·630	3370·	909
910	3370·370	3370·741	3371·111	3371·481	3371·852	3372·222	3372·593	3372·963	3373·333	3373·704	910
911	3374·074	3374·444	3374·815	3375·185	3375·556	3375·926	3376·296	3376·667	3377·037	3377·407	911
912	3377·778	3378·148	3378·519	3378·889	3379·259	3379·630	3380·	3380·370	3380·741	3381·111	912
913	3381·481	3381·852	3382·222	3382·593	3382·963	3383·333	3383·704	3384·074	3384·444	3384·815	913
914	3385·185	3385·556	3385·926	3386·296	3386·667	3387·037	3387·407	3387·778	3388·148	3388·519	914
915	3388·889	3389·259	3389·630	3390·	3390·370	3390·741	3391·111	3391·481	3391·852	3392·222	915
916	3392·593	3392·963	3393·333	3393·704	3394·074	3394·444	3394·815	3395·185	3395·556	3395·926	916
917	3396·296	3396·667	3397·037	3397·407	3397·778	3398·148	3398·519	3398·889	3399·259	3399·630	917
918	3400·	3400·370	3400·741	3401·111	3401·481	3401·852	3402·222	3402·593	3402·963	3403·333	918
919	3403·704	3404·074	3404·444	3404·815	3405·185	3405·556	3405·926	3406·296	3406·667	3407·037	919
920	3407·407	3407·778	3408·148	3408·519	3408·889	3409·259	3409·630	3410·	3410·370	3410·741	920
921	3411·111	3411·481	3411·852	3412·222	3412·593	3412·963	3413·333	3413·704	3414·074	3414·444	921
922	3414·815	3415·185	3415·556	3415·926	3416·296	3416·667	3417·037	3417·407	3417·778	3418·148	922
923	3418·519	3418·889	3419·259	3419·630	3420·	3420·370	3420·741	3421·111	3421·481	3421·852	923
924	3422·222	3422·593	3422·963	3423·333	3423·704	3424·074	3424·444	3424·815	3425·185	3425·556	924
925	3425·926	3426·296	3426·667	3427·037	3427·407	3427·778	3428·148	3428·519	3428·889	3429·259	925
926	3429·630	3430·	3430·370	3430·741	3431·111	3431·481	3431·852	3432·222	3432·593	3432·963	926
927	3433·333	3433·704	3434·074	3434·444	3434·815	3435·185	3435·556	3435·926	3436·296	3436·667	927
928	3437·037	3437·407	3437·778	3438·148	3438·519	3438·889	3439·259	3439·630	3440·	3440·370	928
929	3440·741	3441·111	3441·481	3441·852	3442·222	3442·593	3442·963	3443·333	3443·704	3444·074	929
930	3444·444	3444·815	3445·185	3445·556	3445·926	3446·296	3446·667	3447·037	3447·407	3447·778	930
931	3448·148	3448·519	3448·889	3449·259	3449·630	3450·	3450·370	3450·741	3451·111	3451·481	931
932	3451·852	3452·222	3452·593	3452·963	3453·333	3453·704	3454·074	3454·444	3454·815	3455·185	932
933	3455·556	3455·926	3456·296	3456·667	3457·037	3457·407	3457·778	3458·148	3458·519	3458·889	933
934	3459·259	3459·630	3460·	3460·370	3460·741	3461·111	3461·481	3461·852	3462·222	3462·593	934
935	3462·963	3463·333	3463·704	3464·074	3464·444	3464·815	3465·185	3465·556	3465·926	3466·296	935
936	3466·667	3467·037	3467·407	3467·778	3468·148	3468·519	3468·889	3469·259	3469·630	3470·	936
937	3470·370	3470·741	3471·111	3471·481	3471·852	3472·222	3472·593	3472·963	3473·333	3473·704	937
938	3474·074	3474·444	3474·815	3475·185	3475·556	3475·926	3476·296	3476·667	3477·037	3477·407	938
939	3477·778	3478·148	3478·519	3478·889	3479·259	3479·630	3480·	3480·370	3480·741	3481·111	939
940	3481·481	3481·852	3482·222	3482·593	3482·963	3483·333	3483·704	3484·074	3484·444	3484·815	940
941	3485·185	3485·556	3485·926	3486·296	3486·667	3487·037	3487·407	3487·778	3488·148	3488·519	941
942	3488·889	3489·259	3489·630	3490·	3490·370	3490·741	3491·111	3491·481	3491·852	3492·222	942
943	3492·593	3492·963	3493·333	3493·704	3494·074	3494·444	3494·815	3495·185	3495·556	3495·926	943
944	3496·296	3496·667	3497·037	3497·407	3497·778	3498·148	3498·519	3498·889	3499·259	3499·630	944
945	3500·	3500·370	3500·741	3501·111	3501·481	3501·852	3502·222	3502·593	3502·963	3503·333	945
946	3503·704	3504·074	3504·444	3504·815	3505·185	3505·556	3505·926	3506·296	3506·667	3507·037	946
947	3507·407	3507·778	3508·148	3508·519	3508·889	3509·259	3509·630	3510·	3510·370	3510·741	947
948	3511·111	3511·481	3511·852	3512·222	3512·593	3512·963	3513·333	3513·704	3514·074	3514·444	948
949	3514·815	3515·185	3515·556	3515·926	3516·296	3516·667	3517·037	3517·407	3517·778	3518·148	949
950	3518·519	3518·889	3519·259	3519·630	3520·	3520·370	3520·741	3521·111	3521·481	3521·852	950
951	3522·222	3522·593	3522·963	3523·333	3523·704	3524·074	3524·444	3524·815	3525·185	3525·556	951
952	3525·926	3526·296	3526·667	3527·037	3527·407	3527·778	3528·148	3528·519	3528·889	3529·259	952
953	3529·630	3530·	3530·370	3530·741	3531·111	3531·481	3531·852	3532·222	3532·593	3532·963	953
954	3533·333	3533·704	3534·074	3534·444	3534·815	3535·185	3535·556	3535·926	3536·296	3536·667	954
955	3537·037	3537·407	3537·778	3538·148	3538·519	3538·889	3539·259	3539·630	3540·	3540·370	955
956	3540·741	3541·111	3541·481	3541·852	3542·222	3542·593	3542·963	3543·333	3543·704	3544·074	956
957	3544·444	3544·815	3545·185	3545·556	3545·926	3546·296	3546·667	3547·037	3547·407	3547·778	957
958	3548·148	3548·519	3548·889	3549·259	3549·630	3550·	3550·370	3550·741	3551·111	3551·481	958
959	3551·852	3552·222	3552·593	3552·963	3553·333	3553·704	3554·074	3554·444	3554·815	3555·185	959
960	3555·556	3555·926	3556·296	3556·667	3557·037	3557·407	3557·778	3558·148	3558·519	3558·889	960
M.A.	·0	·1	·2	·3	·4	·5	·6	·7	·8	·9	M.A.

MEAN AREAS 901 to 960.

CUBIC YARDS TO MEAN AREAS FOR 100 FEET IN LENGTH.

M.A.	·0	·1	·2	·3	·4	·5	·6	·7	·8	·9	M.A.
961											961
962											962
963											963
964											964
965											965
966											966
967											967
968											968
969											969
970											970
971											971
972											972
973											973
974											974
975											975
976											976
977											977
978											978
979											979
980											980
981											981
982											982
983											983
984											984
985											985
986											986
987											987
988											988
989											989
990											990
991											991
992											992
993											993
994											994
995											995
996											996
997											997
998											998
999											999
1000											1000
M.A.	·0	·1	·2	·3	·4	·5	·6	·7	·8	·9	M.A.

MEAN AREAS 961 to 1000.

NOTE.—This Table having been carefully computed by the Author, through the usual method of successive additions, and verified in the manuscript, was set up by a skilful printer, and the proofs examined, and re-examined, until they were thought to be free from error; finally, the plates were cast, and the revises taken from them submitted, page by page, to the scrutiny of a competent Civil Engineer, who examined the whole, figure by figure, and ultimately reported but few slight mistakes, which were immediately corrected in the plates themselves; so that every precaution having been taken to secure accuracy:—the Author feels justified in declaring his belief, *that the Table above is entirely clear of any material error.*

SCIENTIFIC BOOKS,

PUBLISHED BY

D. VAN NOSTRAND,

23 MURRAY STREET, and 27 WARREN STREET,

NEW YORK.

———————

PLATTNER'S MANUAL OF QUALITATIVE AND QUANTI-
TATIVE ANALYSIS WITH THE BLOWPIPE. From the last
German edition, revised and enlarged. By Professor TH. RICHTER.
Translated by Prof. HENRY B. CORNWALL, E. M. Illustrated with
eighty-seven woodcuts and one lithographic plate. 560 pages, 8vo.
Cloth. $7.50.

LOWELL HYDRAULIC EXPERIMENTS—being a selection from
Experiments on Hydraulic Motors, on the Flow of Water over Weirs,
and in Open Canals of Uniform Rectangular Section, made at Lowell,
Mass. By J. B. FRANCIS, Civil Engineer. Third edition, revised
and enlarged, including many New Experiments on Gauging Water
in Open Canals, and on the Flow through Submerged Orifices and Di-
verging Tubes. With 23 copperplates, beautifully engraved, and about
100 new pages of text. 1 vol., 4to. Cloth. $15.

FRANCIS (J. B.) ON THE STRENGTH OF CAST-IRON PIL-
LARS, with Tables for the use of Engineers, Architects and Build-
ers. 8vo. Cloth. $2.

USEFUL INFORMATION FOR RAILWAY MEN. Compiled by
W. G. HAMILTON, Engineer. Fourth edition, revised and enlarged.
570 pages. Pocket form. Morocco, gilt. $2.

WEISBACH'S MECHANICS. New and revised edition. A Manual
of the Mechanics of Engineering, and of the Construction of Ma-
chines. By JULIUS WEISBACH, PH. D. Translated from the fourth
augmented and improved German edition, by ECKLEY B. COXE, A. M.,
Mining Engineer. Vol. I.—Theoretical Mechanics. 1 vol. 8vo,
1100 pages, and 902 woodcut illustrations, printed from electrotype
copies of those of the best German edition. $10.

ABSTRACT OF CONTENTS.—Introduction to the Calculus—The General Principles of
Mechanics—Phoronomics, or the Purely Mathematical Theory of Motion—Mechanics,
or the General Physical Theory of Motion—Statics of Rigid Bodies—The Application
of Statics to Elasticity and Strength—Dynamics of Rigid Bodies—Statics of Fluids—
Dynamics of Fluids—The Theory of Oscillation, etc.

"The present edition is an entirely new work, greatly extended and very much improved. It
forms a text-book which must find its way into the hands not only of every student, but of every
engineer who desires to refresh his memory or acquire clear ideas on doubtful points."—The Techno-
logist.

1

A TREATISE ON THE PRINCIPLES AND PRACTICE OF LEVELLING. Showing its application to purposes of Railway Engineering and the Construction of Roads, etc. By FREDERICK W. SIMMS, C. E. From the fifth London edition, revised and corrected, with the addition of Mr. Law's Practical examples for Setting Out Railway Curves. Illustrated with three lithographic plates and numerous woodcuts. 8vo. Cloth. $2.50.

PRACTICAL TREATISE ON LIMES, HYDRAULIC CEMENTS, AND MORTARS. Containing reports of numerous experiments conducted in New York City, during the years 1858 to 1861, inclusive. By Q. A. GILLMORE, Brig.-General U. S. Volunteers, and Major U. S. Corps of Engineers. With numerous illustrations. Third edition. 8vo. Cloth. $4.

COIGNET BETON, and other Artificial Stone. By Q. A. GILLMORE, Major U. S. Corps Engineers. Nine plates and views. 8vo. Cloth. $2.50.

A TREATISE ON ROLL TURNING for the Manufacture of Iron. By PETER TUNNER. Translated and adapted. By JOHN B. PEASE, of the Pennsylvania Steel Works. With numerous engravings and woodcuts. 1 vol. 8vo, with 1 vol. folio of plates. Cloth. $10.

MODERN PRACTICE OF THE ELECTRIC TELEGRAPH. A Hand Book for Electricians and Operators. By FRANK L. POPE. Fifth edition. Revised and enlarged, and fully illustrated. 8vo. Cloth. $2.

TREATISE ON OPTICS ; or, Light and Sight Theoretically and Practically Treated, with the application to Fine Art and Industrial Pursuits. By E. NUGENT. With 103 illustrations. 12mo. Cloth. $2.

THE BLOW-PIPE. A System of Instruction in its practical use, being a graduated course of Analysis for the use of students, and all those engaged in the Examination of Metallic Combinations. Second edition, with an appendix and a copious index. By Professor GEORGE W. PLYMPTON, of the Polytechnic Institute, Brooklyn. 12mo. Cloth. $2.

KEY TO THE SOLAR COMPASS and Surveyor's Companion ; comprising all the Rules necessary for use in the Field. By W. A. BURT, U. S. Deputy Surveyor. Second edition. Pocket Book form. Tucks. $2.50.

MECHANIC'S TOOL BOOK, with practical rules and suggestions, for the use of Machinists, Iron Workers, and others. By W. B. HARRISON, associated editor of the "American Artisan." Illustrated with 44 engravings. 12mo. Cloth. $2.50.

MECHANICAL DRAWING. A Text-Book of Geometrical Drawing for the use of Mechanics and Schools, in which the Definitions and Rules of Geometry are familiarly explained ; the Practical Problems are arranged, from the most simple to the more complex, and in their description technicalities are avoided as much as possible. With illustrations for Drawing Plans, Sections, and Elevations of Buildings and Machinery ; an Introduction to Isometrical Drawing, and an Essay on Linear Perspective and Shadows. Illustrated with over 200 diagrams engraved on steel. By WM. MINIFIE, Architect. Eighth edition. With an appendix on the Theory and Application of Colors. 1 vol., 8vo. Cloth. $4.

2

A TEXT BOOK OF GEOMETRICAL DRAWING. Abridged from the octavo edition, for the use of Schools. By W. MIMFIE. Illustrated with 48 steel plates. New edition, enlarged. 1 vol., 12mo. Cloth. $2.

"It is well adapted as a text-book of drawing to be used in our High Schools and Academies where this useful branch of the fine arts has been hitherto too much neglected."—*Boston Journal.*

TREATISE ON THE METALLURGY OF IRON. Containing outlines of the History of Iron manufacture, methods of Assay, and analysis of Iron Ore, processes of manufacture of Iron and Steel, etc., etc. By H. BAUERMAN. First American edition. Revised and enlarged, with an appendix on the Martin Process for making Steel, from the report of Abram S. Hewitt. Illustrated with numerous wood engravings. 12mo. Cloth. $2.50.

IRON TRUSS BRIDGES FOR RAILROADS. The Method of Calculating Strains in Trusses, with a careful comparison of the most prominent Trusses, in reference to economy in combination, etc., etc. By Brevet Colonel WILLIAM E. MERRILL, U. S. A., Major Corps of Engineers. With illustrations. 4to. Cloth. $5.

THE KANSAS CITY BRIDGE. With an account of the Regimen of the Missouri River, and a description of the Methods used for Founding in that River. By O. CHANUTE, Chief Engineer, and GEORGE MORISON, Assistant Engineer. Illustrated with five lithographic views and 12 plates of plans. 4to. Cloth. $6.

CLARKE (T. C.) Description of the Iron Railway Bridge across the Mississippi River at Quincy, Illinois. By THOMAS CURTIS CLARKE, Chief Engineer. Illustrated with numerous lithographed plans. 1 vol., 4to. Cloth. $7.50.

AUCHINCLOSS. Application of the Slide Valve and Link Motion to Stationary, Portable, Locomotive, and Marine Engines, with new and simple methods for proportioning the parts. By WILLIAM S. AUCHINCLOSS, Civil and Mechanical Engineer. Designed as a handbook for Mechanical Engineers, Master Mechanics, Draughtsmen, and Students of Steam Engineering. All dimensions of the valve are found with the greatest ease by means of a PRINTED SCALE, and proportions of the link determined *without* the assistance of a model. Illustrated by 37 woodcuts and 21 lithographic plates, together with a copperplate engraving of the Travel Scale. 1 vol., 8vo. Cloth. $3.

GLYNN ON THE POWER OF WATER, as applied to drive Flour Mills and to give motion to Turbines and other Hydrostatic Engines. By JOSEPH GLYNN, F. R. S. Third edition, revised and enlarged, with numerous illustrations. 12mo. Cloth. $1.25.

HUMBER'S STRAINS IN GIRDERS. A Handy Book for the Calculation of Strains in Girders and Similar Structures, and their Strength, consisting of Formulæ and Corresponding Diagrams, with numerous details for practical application. By WILLIAM HUMBER. 1 vol., 18 mo. Fully illustrated. Cloth. $2.50.

*** Any of the above Books sent free, by mail, on receipt of price.

☞ My new Catalogue of American and Foreign Scientific Books, 72 pages, 8vo., sent to any address on the receipt of *Ten* cents.

3

www.ingramcontent.com/pod-product-compliance
Lightning Source LLC
Chambersburg PA
CBHW021800190326
41518CB00007B/390